Jhonatan Peres de Sousa

Flow corrosion and corrosion bioinhibitors

Jhonatan Peres de Sousa

Flow corrosion and corrosion bioinhibitors

A theoretical and experimental approach aimed at developing effective prevention and mitigation strategies

ScienciaScripts

Imprint

Cover image: www.ingimage.com

This book is a translation from the original published under ISBN 978-620-6-75795-5.

Publisher:
Sciencia Scripts
is a trademark of
Dodo Books Indian Ocean Ltd. and OmniScriptum S.R.L publishing group

120 High Road, East Finchley, London, N2 9ED, United Kingdom
Str. Armeneasca 28/1, office 1, Chisinau MD-2012, Republic of Moldova, Europe
Printed at: see last page
ISBN: 978-620-8-30317-4

Contents

DEDICATION

I dedicate this work to all those who contributed to this achievement, especially my family.

AUTHOR'S VITAE

Bachelor's degree in Mechanical Engineering from UEMA (2014)
Training in Didactics and Instructional Design at FAMESP (2018)
Pedagogical Training for Non-IFMA Graduates (2022)
Specialisation in Teaching Professional and Technological Education - IFMA (2024)
Master's in Materials Engineering from IFMA (2019)

"Never has man invented anything simpler or more beautiful than a manifestation of nature. Given the cause, nature produces the effect in the shortest way possible."

Leonardo da Vinci

ACKNOWLEDGEMENTS
To GOD, my safe harbour, for giving me all the necessary strength, courage, perseverance and ability to get this far and never give up, even in the midst of the great difficulties I had to face along the way.
To my dear family, especially my mother Evandeny Peres, my father Domingos Lopes and my grandmother Raimunda Peres. Thank you for your love and support during the most difficult times of my life.
To my supervisor Prof Dr Gedeon Reis (IFMA) and co-supervisors Prof Dr Carmen Celia (IFMA) and Prof Dr Maria Eliziane (UFMA) for the opportunity, patience and encouragement in carrying out the work entrusted to me.
To Professor Jose Marcos of NETCOM for his technical support and for sharing information that was fundamental to the realisation of this work.
To Professor Ferdinando Borges of the Department of Mechanics and Materials (IFMA) for his contributions to the work generated.
To all the professors of the Postgraduate Programme in Materials Engineering - PPGEM/IFMA, especially Prof. Dr. Valdemar Silva and Waldemir dos Passos Martins for their advice during the development of this work.
To engineer Renato Conde, for sharing his experience in carrying out electrochemical tests at LABMCD/IFMA. Your help was essential!
METROHM BRASIL for the opportunity to take part in specific training on the Impedance module.
The UFMA ANALYTICAL CENTRE for carrying out the spectrophotometric analyses (FTIR) of various natural extracts.
To my friends at the State University of Maranhao, Jose Carlos, Joao Carlos and Thiago Aguiar, for the lakes created and strengthened, for the conversations and advice that have been essential to my academic life.
To my friend Helismar Medeiros for all his prayers and to the entire IBT-CBN family.
To FAPEMA for the grant. My sincere thanks to everyone!

SUMMARY

In systems that operate with turbulent flow, the corrosive process is accelerated and the study of the application of natural extracts in this condition has become even more feasible, since it simulates conditions that are representative of industrial reality. The aim of this work is to present a theoretical and experimental approach to the application of extracts tested as corrosion bioinhibitors in this flow condition. An experimental platform with electrokinetic flow control was built to carry out the tests. Although the results of some previous evaluations of other extracts tested are presented (palm kernel, babagu and mint), the application of Limonene extract in different concentrations to ABNT 1020 concrete, using a turbulent flow of 3% NaCl, became the focus of this work, as it enabled a more detailed investigation of the phenomena at work. To this end, a number of procedures were carried out, such as: characterisation of the electrolytic flow, electrochemical techniques (OCP, EIE, PPL), optical microscopy and FTIR of the extract. The results concluded that Limonene extract acts as a mixed corrosion inhibitor, reducing corrosion rates and increasing charge transfer resistances at all the concentrations studied. The MO and FTIR evaluations corroborate and complement the electrochemical analyses.

1 INTRODUCTION

Corrosive processes cause a great deal of damage, especially in the industrial sector [1]. Pipelines, oil platforms, metal structures, pipework, bridges and machine components are constantly suffering from this degradation process, causing equipment failures, production line stoppages and high maintenance costs.

When it comes to corrosion, the best way to combat this problem is prevention. Among the various ways of avoiding or minimising the damage caused by metal corrosion, corrosion inhibitors are one of the best and most well-known techniques used by industry, since this method is characterised by its low cost [2]. The acceptance of corrosion inhibitors in industry is justified by their excellent anti-corrosion properties. Inhibitors are already present in many industrial systems, such as refining units, pipework, cooling systems, boilers and oil and gas production processes [3,4]. However, many inhibitors are aggressive to the environment because they are highly toxic due to the presence of chromates and phosphates, and have therefore received strong attention from international authorities.

In this environment, the scientific community has felt the need to investigate environmentally friendly inhibitors called natural inhibitors or green inhibitors [5,6]. These inhibitors consist of natural products such as essential oils and plant extracts that can be used as corrosion inhibitors [7-9].

Therefore, a lot of research has been carried out using green corrosion inhibitors in order to combat this oxidation process in metallic materials. The government has invested in research aimed at discovering inhibitors in water and concrete to protect metals [10]. Other research has studied the effectiveness of these inhibitors in conditions similar to their industrial application, such as systems involving turbulent flow. Studies for this purpose are of fundamental importance, since the corrosion of auger is accelerated, among other factors, by the flow of the electrolyte [11].

The northeastern region is rich in the production of plant extracts and oils, which in turn constitute an important source of income for thousands of families in the state of Maranhão and have a strong potential for application in various industrial segments. Among a series of extracts and oils, Limonene stands out. Extracted from the peel of citrus fruits such as oranges and limes, it is a very versatile compound and there are several applications for it. It is used as a solvent for resins, as a synthesis agent for other chemical compounds, as a dispersing agent for oil, for rubber, paints and various other applications [12].

The purpose of this work is to verify the effectiveness of orange limonene extract as a green corrosion inhibitor in a system with turbulent flow, in order to ensure that the test conditions are representative of cases found in industrial practice. The aim is to assess the corrosion susceptibility of ABNT 1020 steel subjected to a turbulent flow of 3% NaCl in the absence and presence of various concentrations of the extract using electrochemical techniques (OCP, Electrochemical Impedance Spectroscopy and Potentiodynamic Polarisation), Optical Microscopy and FTIR of the extract.

2 JUSTIFICATION

Corrosion is responsible for major losses in industry, especially in processes exposed to aggressive media and influenced by various factors such as flow, temperature, pH, etc. In the oil and gas industry, corrosion represents a large part of the production costs of these companies.

In these cases, when appropriate ways of controlling these corrosion phenomena are not employed, the economic losses can be very high, due to the drop in production resulting from the plant being paralysed. In addition, by controlling corrosion, serious accidents can be avoided that lead to serious social and environmental impacts, including the loss of human life.

Research shows that in Brazil the estimated cost of losses due to corrosion is 3.5% of GDP [13], and in 2016 GDP was R$6.266 trillion, which means that the costs caused by corrosion amount to R$219 billion [14].

Among the various methods used to prevent or control corrosion, the use of inhibitors has proved effective in industry. Their use is due to their low cost and practicality. Although it is effective, some drawbacks such as the generation of waste that causes environmental damage have prompted the scientific community to start looking for new inhibitors with interesting protective characteristics that are environmentally friendly.

The development of this work is justified, firstly, by the need for new inhibitors and, secondly, by the knowledge of their efficiency when subjected to a system with turbulent flow. The work addresses a current and extremely relevant issue, and is therefore an indication of contributions to the scientific community. The novelty of the work is the application of limonene extract as a green corrosion inhibitor for low-carbon auger in a turbulent flow system, a test condition that is representative of the vast majority of industrial pipework.

3 OBJECTIVES

1.1 General objective

Investigate the effectiveness of Limonene extract as a green corrosion inhibitor for carbon bag in a turbulent flow system.

1.2 Specific objectives

- To build and evaluate the functionality/sensitivity of an experimental bench coupled to an electrochemical cell that simulates a system with turbulent flow and allows the speed of the electroKtic solution (3% NaCl) to be controlled;
- Evaluate the interaction of the Limonene extract in different concentrations in a turbulent flow of 3% NaCl at 0.62 $m.s^{-1}$ (2.91 L/min) or 6583.58 Reynolds;
- Calculate the efficiencies for the various concentrates and characterise the interface of the ABNT 1020/NaCl 3% auger in the absence and presence of the Limonene extract in various concentrates by means of Electrochemical Impedance (EIE) in the midst of the turbulent flow of NaCl 3% at 0.62 m.s $;^{-1}$
- Evaluate the efficiency of the extract in the presence of Limonene in various concentrations by means of Potentiodynamic Polarisation tests under the same flow conditions;
- Determine the current density and corrosion potential, cathodic and anodic coefficients in the absence and presence of the extract at various concentrations with a flow of 0.62 $m.s^{-1}$ using the Tafel Extrapolation method;
- Discuss the action of the extract, present the adsorption isotherm and propose an inhibition mechanism for the case studied;
- Check the preservation of the sample after prolonged tests by microscopy;
- Using Fourier Transform Infrared Spectroscopy (FTIR), identify the presence of functional groups present in Limonene that influence the extract's action as a corrosion inhibitor.

4 LITERATURE REVIEW

4.1 Metal corrosion

According to Gentil [15], corrosion is a spontaneous process linked to the deterioration of a material, usually metallic, as a result of chemical or electrochemical action by the environment, which may or may not be associated with mechanical forces. The result of this process is the weakening of the material, as it causes losses in its essential qualities, such as mechanical strength, elasticity and ductility, since the product of corrosion is extremely poor in terms of these properties [16].

Corrosion is best understood based on the mechanisms of corrosive processes. These processes are always associated with several variables that act as indicators for the correct choice of material to be used in certain equipment, installations, etc. The variables are [15]:

- **Metallic material** - chemical composition, production process, heat and mechanical treatments, surface condition, shape, joining of materials (welding, among others), contact with other metals and presence of impurities;
- **Corrosive medium** - chemical composition, concentration, impurities, pH, temperature, oxygen content, pressure and suspended solids;
- **Operating conditions** - mechanical stresses, relative movement between metal material and medium, conditions of immersion in the medium (total or partial) and means of protection against corrosion.

4.1.1 Corrosion mechanisms

Corrosion can occur by chemical or electrochemical action; however, electrochemical action is observed more frequently. Some aspects distinguish between these two types of corrosion [15]:

- **Chemical** corrosion - This process takes place in the absence of H_2O at high temperatures and involves direct contact between the metal and the corrosive medium. This type of action is common in equipment that operates in high temperatures, such as ovens, boilers, etc.
- **Electrochemical action** - This process is similar to how batteries work. A flow of electrons from one metal to the other originates from the difference in electrical potential between the two metals, which are interconnected in an electrolytic medium that acts as an ionic conductor.

The electrochemical action illustrated in Figure 1 can be observed in three main stages: anodic process, displacement of electrons and tones and cathodic process. In the anodic process, the tones are transferred to the solution. The displacement of electrons and tones occurs from anodic regions to cathodic regions by means of the metallic circuit and also by the diffusion of anions and cations in the solution and in the cathodic process the reception of electrons by the tones or molecules in the electrolyte occurs [15].

Figure 1 shows a schematic of electrochemical action:

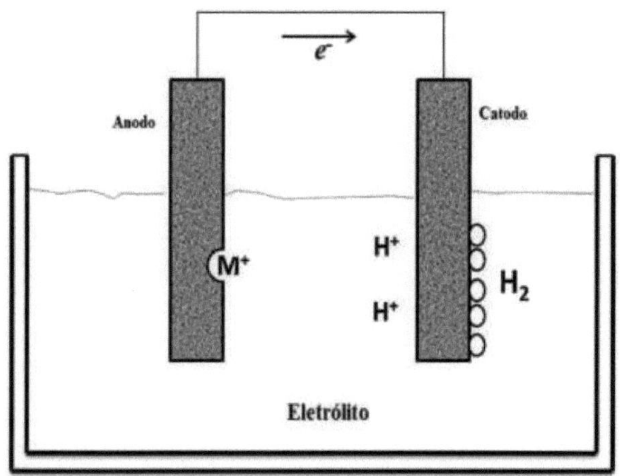

Figure 1: Schematic of electrochemical action. Adapted from Gentil [15].

Faraday's Law presents a number of ways of knowing the intensity of the corrosion process. Its formulations describe that the intensity of the corrosion process can be evaluated by the number of charges of the tones that pass into solution at the anode, by the number of charges of the tones discharged at the cathode or even by the number of electrons that migrate from the anode to the cathode [15].

Carbon auger pipes have reactions that have been described by Gentil [15] in Equations 4.1 to 4.5:

- Anodic area:

$$Fe \rightarrow Fe^{2+} + 2e \qquad (4.1)$$

Cathodic area (non-aerated medium):

$$2\,H_2O + 2e \rightarrow H_2 + 2\,OH^- \qquad (4.2)$$

Cathodic area (aerated environment):

$$H_2O + \tfrac{1}{2}\,O_2 + 2e \rightarrow 2\,OH^- \qquad (4.3)$$

The corrosion product may appear as:

$$Fe^{2+} + 2\,OH^- \rightarrow Fe(OH)_2 \qquad (4.4)$$

$$3\,Fe(OH)_2 \rightarrow Fe_3O_4 + 2\,H_2O + H_2 \qquad (4.5)$$

4.1.2 Forms of corrosion

There are various classifications of corrosion. They vary according to the medium, the cause of the attack, the corrosion product, the mechanical factors, the location of the attack and the kinetics present in the process [15]. The main techniques registered with the American Society of Metals (ASM) for identifying corrosion include visual inspection, metallographic examination, mass loss, fire depth measurement and non-destructive inspection [16,17].

According to Rossi [18], corrosion is classified as chemical or electrochemical. In more detail, it can also be divided according to its morphology (uniform or localised - alveolar or pyiform, selective, intergranular and transgranular) and also according to its phenomenology (galvanic, through gaps or under deposition - occluded cell, erosion, fatigue, under tension, atmospheric and microbiological).

Gentil [15] presents the main types of corrosion found in oil pipelines (see Table 1) and that in some cases, according to Rossi [18], these processes are triggered by high levels of salinity.

Table 1: Main types of corrosion found in oil pipelines. Adapted from Gentil [15].

TYPE	MORPHOLOGY	DESCRIPTION
ALVEOLAR		This is the type of corrosion in which the entire surface exposed to the medium is corroded uniformly, resulting in a gradual and uniform decrease in the thickness of the section. It is also known as generalised corrosion and can be seen in carbon steel pipes.
UNIFORM	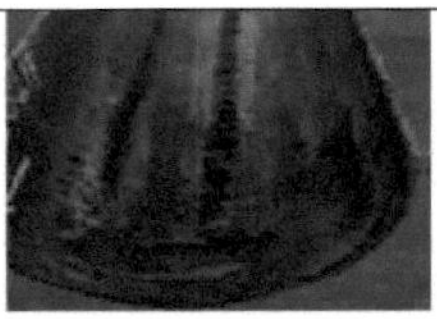	This type of corrosion develops on the metal surface, producing cavities similar to alveoli characterised by a rounded profile and a bottom that is generally smaller than the diameter.
PITIFORM	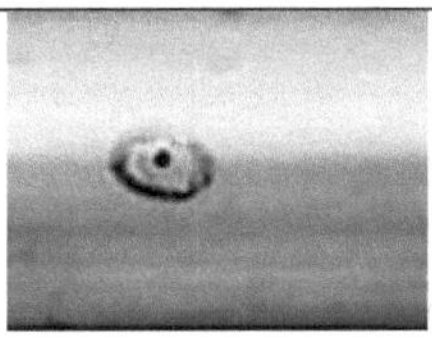	It is a type of corrosion that is extremely punctual on the surface of the metal. These locations are called pits, which are cavities that have an angular depth and are generally larger than their diameter. They are not common in carbon steel, but are more common in stainless steels.

4.1.3 Chloride corrosion

Most of the problems in industrial installations involving corrosion are caused by brines. An example of this is oil pipelines in the oil industry, where the corrosion process is initiated by high salinity levels. Although buried pipelines have internal and external coatings, they are not exempt from the effects of corrosion. The decrease in the pipe's capacity to withstand external loads, such as those resulting from traffic loads, thermal loads and internal pressure are just a few of the effects worth highlighting. In general, the magnitude of the risks caused by chloride corrosion reach scales that deserve full attention [19].

According to Anjos [20], salt water consists of a relatively uniform solution made up mainly of sodium and magnesium chlorides dissolved in water, which also contains small amounts of other minerals and can be considered equivalent to a 0.5 M solution of sodium chloride. The chloride tones present in this type of solution are extremely aggressive, as they damage the passive layer of the auger, triggering a process of embrittlement and subsequently material failure.

Although there is no consensus in the scientific community on the depassivation of auger by chlorides, the ACI 222 Manual [21] presents three theories that explain the mechanism of action of chlorides in the corrosion of auger:

- **The oxide film theory:** this theory explains that chloride tones penetrate existing defects in the passivating oxide film more easily than other aggressive tones and can disperse in this film allowing the attack and weakening of the auger.
- **The adsorption theory:** this theory states that an adsorption process takes place on the surface of the auger involving chloride tones and hydroxyl tones or chloride tones and dissolved oxygen, promoting the hydration of iron and enabling its dissolution into cations.
- **The transient complex theory:** this theory states that ferrous corrosion compounds called transient complexes are formed from chloride tones (Cl^-) and hydroxyl anions (OH^-). These diffuse from the anode towards the cathode, facilitating the continuation of corrosion.

It should be noted that the oxidation reaction of metallic materials caused by exposure to marine environments is one of the most severe [21] and can be represented by Equation 4.6 below:

$$O_2 + 2\,H_2O + 4e \rightarrow 4\,(OH)^- \qquad (4.6)$$

4.1.4 The use of carbon auger

Currently, the main materials used in the metallurgical industry are steel and cast iron, which essentially includes iron-carbon alloys. Being relatively cheap, carbon (C) is an extremely efficient alloying element for hardening iron. ABNT classifies carbon auger by its carbon content. Between 0.1 and 0.3 per cent carbon, the auger is considered low carbon; between 0.30 and 0.85 per cent, it is classified as medium carbon. If this content exceeds 0.85% and reaches a maximum of 1.5%, it is classified as high carbon.

Among a range of metal alloys, carbon steel is the one with the most favourable combination of properties, making it widely used in engineering applications [22,23]. Due to its versatility in applications, this material is often used in atmospheric exposure conditions and can easily be found in large structures such as bridges, viaducts, large pipes in water treatment plants, machine components and household utensils [24]. Figure 2 illustrates the use of carbon bagasse in the Pedra do Cavalo Adductor System in Simoes/BA.

Figura 2: Carbon steel pipework in the Pedra do Cavalo - Simoes Filho/BA Adductor System [25].

Although this material has numerous applications, the susceptibility to corrosion of carbon auger limits its useful life depending on the conditions to which it is exposed. Even though carbon auger can develop a passive layer of iron oxide, exposure to aggressive environments such as chlorides can lead to a rupture of the passive layer, thus accelerating the corrosion of the vulnerable surface and subsequently forming a layer of passive (oxidised) iron oxide [26].

4.1.5 Corrosion control

The countless structural losses and high maintenance costs in industrial plants caused by the limited resistance of carbon auger in highly corrosive media has fuelled the search for new corrosion control methods. The methods reported by Gentil [15] are shown in Figure 3.

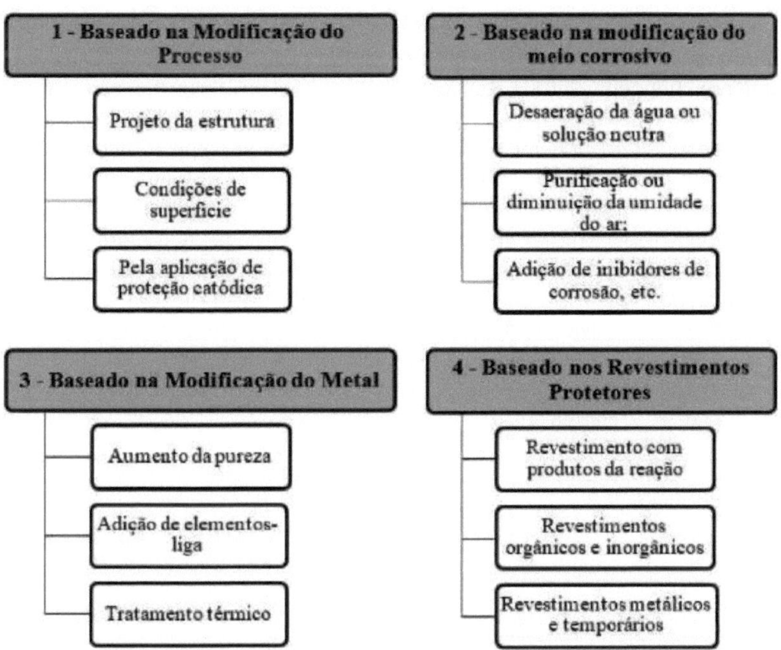

Figura 3: Protection methods for corrosion control. Adapted from Gentil [15].

For all the methods presented, the economic and environmental factor must be taken into account, since a detailed analysis of the costs of application, advantages and disadvantages of using a particular protection is necessary. According to Gentil [15], the costs arising from equipment deterioration and the damage resulting from this deterioration (unit stoppages, loss of efficiency, loss of products and contamination) are factors that must be carefully analysed. In addition, the environmental impacts caused by certain protection methods have received strong attention and concern from international authorities. The application of a particular protection becomes unviable when the value of the environmental decontamination required to correct the environmental impacts caused by certain corporations reaches high values [27].

Thus, with the growing increase in environmental awareness, the method of modifying the corrosive environment by adding corrosion inhibitors has gained prominence, since they are ecologically acceptable compounds, have a low cost and are abundant in nature [28].

4.2 Adsorption of molecules on metal surfaces

By definition, adsorption is a mass transfer operation related to the ability of certain solids to concentrate substances present in liquid or gaseous fluids on their surface through spontaneous interaction under the action of attractive forces. The literature classifies adsorption based on the magnitude of the molecular forces involved into two types: physical adsorption and chemical adsorption [29].

Physical adsorption, also known as physisorption, is well characterised by the relatively weak interactions of Van der Waals forces acting between the adsorbed

species (adsorbate or adsorbate) and the surface of the adsorbent [29]. The magnitude of the forces is similar to molecular cohesive forces and, in general, this phenomenon happens quickly, but can be slow if it involves the occupation of a porous medium [29-31].
Chemical adsorption or chemisorption, unlike physical adsorption, occurs through a mechanism of exchange or sharing of electrons between the adsorbate molecules and the surface of the adsorbent, thus generating a chemical reaction. The new chemical bond resulting from this process is much stronger and more specific than those found in physical adsorption [29,31]. The main differences between physical and chemical adsorption are summarised in Table 2.

Table 2: Difference between physical and chemical adsorption processes [29,31].

Physical Adsorption	Chemical Adsorption
▪ Low chord charge (generally below 10 kcal/mol);	▪ Heat of adsorption above 20 kcal/mol;
▪ Not specified;	▪ Highly specific;
▪ Can be either monolayer as well as in multilayers;	▪ Occurs only in monolayer;
▪ There is no dissociation of the species adsorbed;	▪ May involve dissociation;
▪ No electron transfer , although polarisation can occur;	▪ With electron transfer and formation of a chemical bond between the absorbate and the adsorbent;
▪ Fast, low activation and reversible in non-porous solids.	▪ Activated, can be slow and irreversible.

4.2.1 Langmuir isotherm

In all adsorption processes, the amount of substance adsorbed on the surface of the material decreases with increasing temperature, characterising an exothermic process. However, at constant temperatures, adsorption increases with the concentration of the inhibitor. Adsorption isotherms are diagrams controlled by thermodynamic principles that represent a relationship between the amount adsorbed and the concentration of the inhibitor at a given temperature. These diagrams provide extremely relevant information on adsorption mechanisms [29,32].
In general, adsorption isotherms have the shape of a hyperbole, classifying adsorption as favourable, extremely favourable or unfavourable [29]. The literature presents various models, among which we will highlight the Langmuir adsorption isotherm.
The simplest theoretical model for adsorption in monolayers, which is used in most scientific studies, is the Langmuir isotherm. When describing this model, Langmuir made some considerations such as [29]:

- The adsorbate molecules are chemically adsorbed by a number of well-defined rivers;
- Each sftio corresponds to a single molecule;
- All adsorption systems are energetically equivalent with perfectly flat surfaces on a microscopic scale; and
- The adsorption of a molecule in a given site is independent of the occupation of

the neighbouring sites, i.e. it is a process without interactions between adsorbed molecules.
The Langmuir equation is represented by Equation 4.7.

$$\frac{C}{\theta} = C + \left(\frac{1}{K_{ads}}\right) \quad (4.7)$$

Where C represents the concentration of the inhibitor at equilibrium, θ is the degree of surface coverage and K_{ads} is the adsorption equilibrium constant.
The degree of surface coating can be expressed as a function of the corrosion current according to Equation 4.8.

$$\theta = \frac{i_{st} - i_{ci}}{i_{st}} \quad (4.8)$$

Where i_{si} represents the corrosion current without inhibitor and i_{ci} represents the corrosion current with inhibitor.

4.2.2 Variation of the Gibbs - Helmholtz function

In the study of adsorption, thermodynamic parameters are extremely important because, among other factors, they indicate the nature of the process, thus enabling us to understand the mechanisms at work [29].
The thermodynamic parameters most commonly used in the study of adsorption are the equilibrium constant (K_{ads}) and the Gibbs energy variation (AG°) [28,29]. These parameters are correlated using Equation 4.9.

$$K_{ads} = \frac{1}{C_{solv}} \exp\left(\frac{\Delta G^{o}}{RT}\right) \quad (4.9)$$

In this equation, K_{ads} is the adsorption constant, C_{SOlv} is the solvent concentration in mg/L, R is the universal gas constant (8.3145 $J.K^{-1}.mol^{-1}$), T is the temperature in K and ° is the Gibbs function variation in kJ/mol.
The literature explains that, as transformations occur, the algebraic sign of AG° will indicate in which direction the transformation will occur. The possibilities are described below [29]:

- AG° < 0: the transformation can occur spontaneously or naturally;
- AG° = 0: the system reaches thermodynamic equilibrium; and
- AG° > 0: the system is endergonic and the transformation is non-spontaneous.

4.3 The role of organic and inorganic inhibitors in corrosion

4.3.1 Initial considerations

Corrosion inhibitors can be defined as substances that have the ability to slow down anodic reactions (anodic inhibitors), cathodic reactions (cathodic inhibitors) or both (mixed inhibitors). In other words, they are a substance or mixture of substances that, when present in adequate concentrations in the corrosive environment, reduce or eliminate corrosion [15,33-35].
Corrosion inhibitors are classified in different ways according to the point of view of various researchers. However, the main classification relates to the chemical nature and mechanisms of action. Dariva et al.[3] classifies inhibitors according to Figure 4.

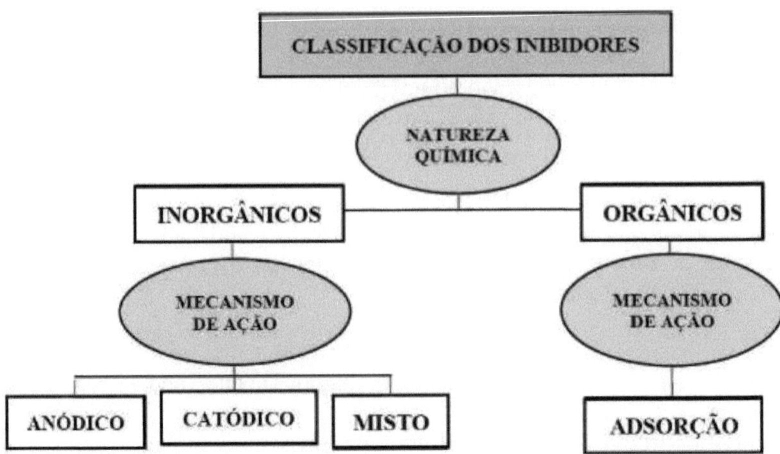

Figure 4: Classification of corrosion inhibitors. Adapted from Dariva [3].

4.3.2 Inorganic inhibitors

Inorganic inhibitors with an anodic effect (also known as passivation inhibitors) act through an anodic reduction reaction, i.e. they slow down or block reactions at the anode. In general, this type of inhibitor reacts with the corrosion product initially formed, giving rise to a cohesive and extremely insoluble molecule on the metal surface, promoting anodic polarisation [3,15]. The anodic inhibitors react with the metal tones produced at the anode, subsequently forming insoluble hydroxides which are deposited on the metal surface as an insoluble molecule impermeable to the metal tones. This process results in OH- tones [15].

The effect of these inhibitors can be observed experimentally, as their presence shifts the corrosion potential towards more positive values while the current value decreases in the curve (as shown in Figure 5A). Some examples of anodic inorganic inhibitors are nitrates, molybdates, sodium chromates, phosphates, silicates and hydroxides.

Inhibitors with a cathodic effect are substances that release metallic tones capable of reacting with cathodic alkalinity, producing insoluble precipitates that act to restrict cathodic reactions. The cathodic regions are covered by this insoluble compound, thus hindering the diffusion of oxygen and the conduction of electrons, which in turn inhibit the cathodic process. This inhibition promotes a marked cathodic polarisation [15].

Zinc, magnesium and nickel sulphates are used as cathodic inhibitors, since the Zn^{2+} , Mg^{2+} and Ni^{2+} tones form insoluble hydroxides with the hydroxyls (OH^-) in the cathodic areas: $Zn(OH)_2$, $Mg(OH)_2$, $Ni(OH)_2$, interrupting or hindering the corrosive process. Polyphosphates, phosphonates, tannins and lignins can also be mentioned, as they present the same reaction mechanism [3,15]. Experimentally, the presence of cathodic inorganic inhibitors shifts the corrosion potential towards more negative values, as illustrated in Figure 5B.

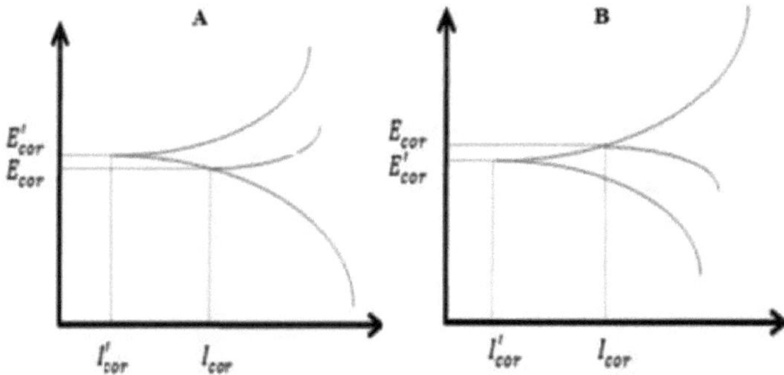

Figura 5: Potentiostatic polarisation diagram: action of the anodic inhibitor (A) and the cathodic inhibitor (B) [3].

The action of mixed inhibitors is generally due to adsorption processes, since they act as protective molecules, they can act on both the metal surface and the passivating molecule, thus forming a continuous, protective layer capable of protecting both cathodic and anodic areas, making them more efficient than their individual effects.

4.3.3 Organic inhibitors

The organic compounds used as corrosion inhibitors can act anodically, cathodically or perform a mixed action [3]. However, in the majority of cases this action takes place through a process of surface adsorption which consists of the formation of a film on the metal surface. The affinity of the molecules with the metal surface promotes the accumulation of a hydrophobic protective molecule that is adsorbed on it, thus providing a barrier to the dissolution of the metal in the presence of the electrolyte. The inhibitors must be soluble and dispersible in the tested medium [15,36].

Figure 6 shows a theoretical Upic potentiostatic polarisation curve representing the effect of the organic inhibitor on anodic and cathodic behaviour. According to the curves, after adding the inhibitor, it can be seen that the current decreases from Icor. to I'cor. while the corrosion potential remains the same [3].

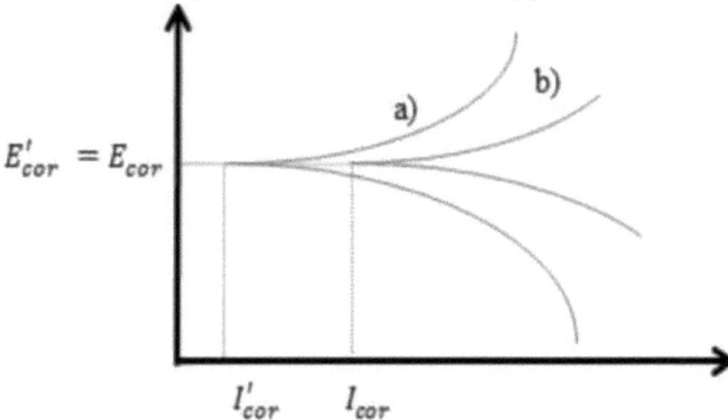

Figura 6: Theoretical potentiostatic polarisation diagram: behaviour of a metal in

a solution with (a) and without (b) corrosion inhibitor [3].

Figure 7 shows the mechanism of action of the organic inhibitors when they undergo the adsorption process, thus giving rise to the thin protective layer made up of Inh inhibitor molecules that become lodged on the metal surface.

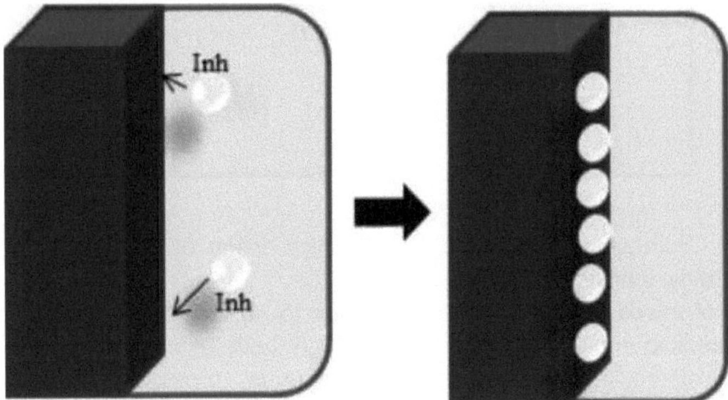

Figura 7: Illustration of the mechanism of action of the organic inhibitor through the adsorption of the inhibitor on the metal surface [3].

According to Sanyal [37], the effectiveness of an organic inhibitor depends on a number of factors:

- Chemical structure, such as the size of the organic molecule;
- Aromaticity and/or conjugated bond, such as the length of the carbon chain;
- Type and number of atoms or bonding groups in the molecule n or o;
- Nature and charges of the metal surface of the adsorption mode, as well as the strength of the metal substrate;
- Ability for the protective layer to become compact or cross-linked; and
- Type of electrolyte solution with adequate solubility in the environment.

The polar function has a major influence and constitutes the centre of the reaction for establishing the adsorption process [36]. Guo et. al [38] explains that the empirical rule relating the efficiencies (O < N < S) is incomprehensible, since oxygen has a higher electronegativity (x_O=3.5) than nitrogen (XN =3.0) and sulphur (x_S=2.5). The correctness of this empirical rule has been verified experimentally by some researchers. Figure 8 illustrates a comparison of the capacity of O, N and S atoms to donate electrons.

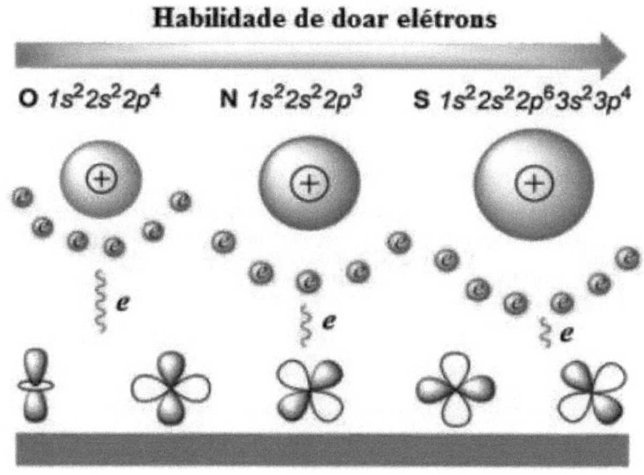

Figura 8: Ability of O, N and S atoms to donate electrons [38].

Research by El-Haddad et. al [39] reports that the adsorption of oxygen, nitrogen and/or sulphur present in the organic inhibitor blocks active corrosion sites on the metal surface. Compounds with n-bonds are the most effective and efficient inhibitors, but they have high biological and environmental toxicity.

The class of organic inhibitors can be subdivided into synthetic organics and natural organics. Synthetic inhibitors consist of artificially produced commercial inhibitors that act in the vapour phase, also known as volatile corrosion inhibitors (VCI). Its action can be explained by the formation of a monomolecular, invisible molecule on the metal surface and its application has been observed in the storage and transport of metal materials through closed packaging in the absence of protective oils or greases [3,15].

Natural organic inhibitors, also known as "green inhibitors", are organic compounds that work by adsorbing tones or molecules onto the metal surface, minimising the corrosion rate by reducing the anodic reaction, reducing the rate of diffusion of the reagents to the metal surface and also the electrical resistance of the metal surface [28].

Natural organic inhibitors include ascorbic acid, succinic acid, tryptamine, caffeine and extracts of natural origin. Many of these inhibitors come from fruit, as they are a rich source of chemicals such as vitamins, minerals and phenolic compounds [36].

4.3.4 Natural organic inhibitors

National and international research shows that plant extracts are becoming increasingly important as an environmentally friendly source. The availability of these extracts in nature, their low cost and the fact that they are renewable for a wide variety of plants are the factors driving the scientific community's search for green inhibitors. According to Grosser [28], nature has a rich source of components that have a high inhibition efficiency.

Extracts derived from leaves, branches, stems, roots and fruits are the target of many researchers and some have been studied to act as effective corrosion inhibitors in different aggressive media. Normally, the inhibiting power of these extracts is attributed to the presence in their composition of complex organic species such as tannins, alkaloids and nitrogenous bases, carbohydrates and proteins [40,41].

The extract of the plant *Myrmecodia Pendans*, known as the ant plant, was studied by Pradityana et al in a 3.5% NaCl solution using a carbon sandwich electrode [42]. Electrochemical measurements confirmed the inhibitory action of this compound, which showed a maximum efficiency of 79.7% (concentration of 400 mg/L). The authors concluded that the system obeys the Langmuir isotherm and that the inhibition mechanism occurs through adsorption processes of the extract on the metal surface forming a thin protective layer [42].

Vaszilcsin et al [43] studied the efficiency of the capsaicin extract from chilli peppers (*Capsicum annuum Group*) as a natural corrosion inhibitor for carbon auger in 3.5% NaCl solutions using electrochemical techniques. It was observed that the inhibition efficiency increased as the concentration of the extract increased, with 20 $mg.L^{-1}$ providing an efficiency of 91.19%. Morphological analyses of the samples submitted to SEM revealed the preservation of the metal surface in the presence of the extracts. The authors attributed the inhibitory effect to the adsorption of capsaicin, an alkaloid compound, on the metal surface.

Yetri et. al [44] investigated the anticorrosive power of cocoa bark extract (*Theobroma cacao*) in combating the corrosion of carbon steel using mass loss, potentiodynamic polarisation and impedance measurements, among others. An efficiency of 90.19% was recorded for the mass loss test, 75.23% for EIE and 90.19% for the polarisation test in the presence of 2.5% V/V extract. These results were justified by the formation of a protective molecule adsorbed on the surface made up of phenols, aromatic rings and ethers identified in the extract.

The inhibition effects of *Curcuma longa* plant extract on the corrosion of carbon auger in 3.5% NaCl + 16 ppm Na_2S was investigated by Fakih et al [45] using electrochemical techniques, among others. The results indicated that the extract acts as a mixed inhibitor with a predominant cathodic effect with an efficiency of 79.4% in the presence of 250 ppm of the extract. Impedance measurements revealed an increase in charge transfer resistance at all concentrations investigated. Graphical analyses suggest the Temkim model as a mechanism that best represents the adsorption process.

Tomie et. al [46] investigated *Salvia officinalis* known in Brazil as Salvia as a green corrosion inhibitor for carbon auger in a 3% NaCl medium at different extract concentrations using gravimetric and electrochemical methods. The authors suggest the extract as an inhibitor in 3% NaCl, since the highest degree of protection was 97.5% obtained in the presence of 1.5 g/dm^3 of the plant extract for 6 hours. The impedance spectra confirmed this result, showing that it is a reliable method for rapid corrosion tests.

Haider et al [47] investigated the inhibitory action of *Cactaceae*, popularly known as Cactus, on carbon agar in 3.5% NaCl in the absence and presence of the extract at various concentrations and temperatures. Potentiodynamic polarisation tests revealed an efficiency of 94% in the presence of 1 ppm of the extract at a temperature of 298 K. The authors concluded that the extract acts as a mixed inhibitor and that the inhibition efficiency is linked to the presence of polyphenolic compounds, more precisely to heterocyclic electrons and atoms.

The extract from the leaves of the Bombardeira tree (*Calotropis procera*) was studied by Ibrahim et. al [48] as a corrosion inhibitor for carbon steel in a 3.5% NaCl solution using electrochemical techniques. The analyses showed a mixed action of the inhibitor with an efficiency of 80% in the presence of 50 ppm of the extract. The presence of N and O in the extract, as well as n-electron systems coupled to high interaction adsorption energies at the extract/carbon-aggregate interface are responsible for the corrosion inhibition performance.

Al-Sultani [49], Jokar [50] and Challouf et. al [51] have also carried out important studies in this field. The investigations carried out by these and the other researchers cited show agreement between the electrochemical techniques and all of them address the process of compound adsorption as fundamental to the extract's action as a corrosion inhibitor. It is also worth pointing out that, in all the studies observed, the efficiency proved to be dependent on the concentration of the extract.

4.4 Effects of electrolyte flow on corrosion

4.4.1 Physical aspects

According to Ferreira [52], the intensity and mechanism of corrosion in a pipeline are strongly dependent on surface phenomena, which include the interactions present in the internal wall interface of the pipeline - electrolyte flow. Therefore, the kinetics of corrosion under flow conditions are totally dependent on the hydrodynamic factors of the medium.

The flow regimes of a fluid inside a duct can be classified as laminar or turbulent, based on its structure. In the laminar regime, the particles move in individualised laminae, grouped together and without any exchange of mass between them, while in the turbulent regime the fluid particles travel in a random direction, i.e. the velocity has components that are transversal to the general movement of the whole [53,54].

The characterisation of these flows follows a physical approach presented by Reynolds [54]. Reynolds found that the type of flow present in pipes can be calculated by an expression called the Reynolds number (Re) described in Equation 4.10.

$$Re = \frac{\rho . v . D}{\mu} = \frac{v . D}{\nu} \qquad (4.10)$$

In the above expression, D represents the diameter of the duct (m), v the average flow velocity (m/s), *v* the kinematic viscosity, p the dynamic viscosity (Kg/m.s) and p the specific mass Kg/m^3). Reynolds also made the following considerations:

- Re < 2,000: Laminar flow;
- 2,000 < Re < 2,400: Transition Flow; and
- Re > 2,400: Turbulent flow.

In practice, if Re > 2,320, the flow is turbulent, otherwise it is always laminar. In industrial flow measurements, the flow regime is in most cases turbulent with Re > 5,000.

4.4.2 Model for the interfacial dynamics of the elements

The phenomenon of the formation and stability of the passive layer of the metal in the midst of a certain flow is strongly influenced by the kinetic speed with which this flow circulates in the system. The interactions between the electrolyte particles increase as the speed of the flow increases, since this provides an increase in the kinetic energy of the species involved in the process when compared to the system in the absence of flow.

To better explain this procedure, Ochoa [55] developed a model (Figure 9A and 9B) capable of illustrating the mechanism of passive layer formation in AISI 304 steel. In general, the species located close to the wall region have enough kinetic energy to affect the tones and modify the structure of the electrical double layer.

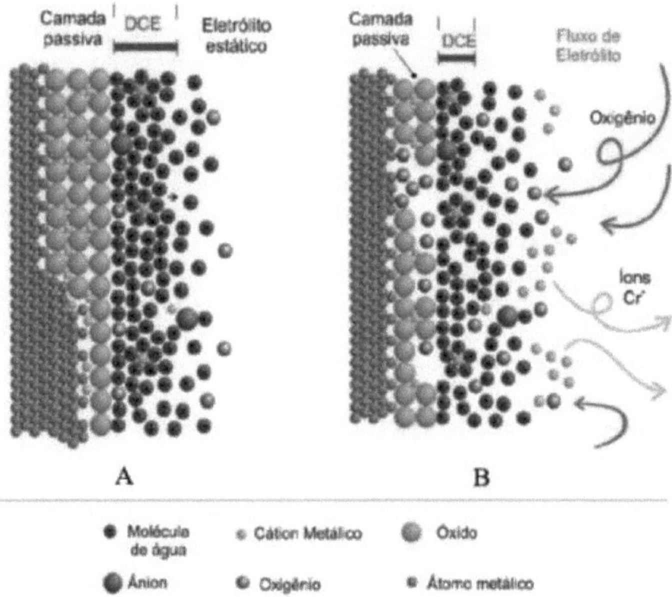

Figure 9: Mechanism of passive layer formation on AISI 304 in 3.5% NaCl static electrolyte (a) and for moving electrolyte (b) [55].

In corrosion reactions, the aggressive tones must reach the metal surface by molecular diffusion as a result of a concentration gradient and the corrosion products must be displaced away from the surface. In this dynamic, the dissolved Fe+ tones move away, increasing the reaction rate [55].

The atoms of the metal surface enter the stagnant solution (Figure 9A) as tones, by means of an electrochemical process. In the formation of the passive layer, the oxygen in flow conditions is facilitated in its movement to the surface, but the Fe+ tones required for the formation of the passive layer are displaced further away as a result of the increase in mass transfer through the diffusion layer, thus hindering the formation of the layer, as shown in Figure 9B. In comparison, the passive layer formed under flow conditions is thinner, more unconfined and less resistant than the one formed under stagnant conditions [55].

4.4.3 Effect of turbulent flow on corrosion

Research carried out in past decades hardly took into account the effect of flow conditions on the efficiency of these inhibitors. However, recent research has

revealed that turbulent flow conditions can influence experimental results and, consequently, the corresponding decisions [56].
Despite the importance of this parameter, there are few scientific studies aimed at testing corrosion inhibitors in flow systems, since there is already a consensus among several researchers that research in this area is fundamental, as it provides information that helps in the choice of efficient materials and inhibitors in conditions involving flow. Interest in research analysing the influence of flow on the corrosive process has been addressed by a number of researchers.
Olvera et. al [56] studied the influence of turbulent flow on the performance of corrosion inhibitors on API 5LX32 auger using the rotating cylinder electrode (RCE) technique in aqueous environments containing CO_2. It was noted that corrosion is influenced by flow and that internal corrosion can be controlled by the addition of inhibitors through film formation on the metal surface. The authors concluded that, at constant speeds, with an increase in concentration there is a decrease in the corrosion rate. It was suggested that there is a limiting concentration, i.e. better performance is not always associated with an increase in inhibitor concentration.
Jing et. al [57] investigated the effect of flow velocity on the corrosion behaviour of AZ91D magnesium alloy in an elbow using a closed loop system combined with polarisation and fluid dynamics techniques.
(CFD). It was found that the corrosion rate increases with increasing flow velocity and that a critical flow velocity could exist in the corrosion of AZ91D magnesium alloy. It was found that fluid flow increases the convection and diffusion of reactive species, accelerating electrochemical corrosion reactions, and that flow accelerates the removal of corrosion products from the electrode surface [57].
Li et al [58] analysed the corrosion of X65 auger in the presence of CO_2 in oil-free and saturated solutions (oil-water emulsions) subjected to continuous flow. It was found that, in both solutions, the mass transfer of corrosive species, such as H^+ and H_2CO_3, dominates the cathodic processes during auger corrosion. The authors concluded that increasing the flow velocity accelerates mass transfer, contributing to increased corrosion. The presence of oil in the fluid decreased auger corrosion at all concentrations, a fact attributed to the increased possibility of the pipe wall becoming wetter when the fluid contains more oil.
Leon et al [59] studied the behaviour of AISI 1018 carbon steel in buffered chloride solution with and without hydrocarbons in a continuous flow system using electrochemical techniques. It was found that the presence of hydrocarbons in the system increases the currents for certain immersion times compared to the blank system and that the impedance values decrease in the absence of hydrocarbons, indicating higher corrosion rates. This effect was attributed to the presence of some hydrocarbon in contact with the metal surface which, in some way, favours the interaction of corrosive species.

4.5 Presentation of Orange Limonene

Limonene, officially classified by IUPAC as 4-isoprenyl-1-methyl-cyclohexene, is a monocyclic monoterpene found in more than 300 plants [60]. With the molecular formula $C_{10}H_{16}$, Limonene has two isomers (S and R) which are the most abundant monoterpenes in nature. The S-(-) isomer is found in a wide variety of plants and herbs such as *Mentha spp,* while the R-(+) isomer is the major component of oils extracted from lemon and orange peels, reaching concentrations of 90 to 96%. Figures 10A and 10B illustrate the chemical structure of the S-(-) and R-(+) isomers respectively and Figure 10C shows the 3D molecular structure.

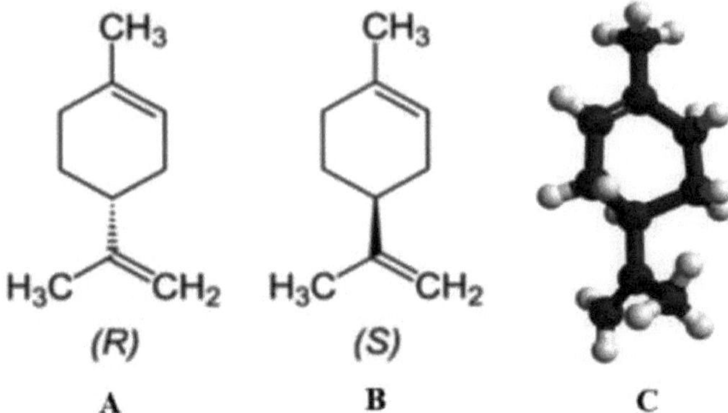

Figure 10: Representation of the R (10A) and S (10B) isomers and 3D structure of the Limonene molecule (10C). Adapted from [61] and [62].

Orange production, the fruit from which Limonene can be extracted, exceeds 80 million tonnes per year, making it one of the most produced and consumed fruits in the world. On average, 34% of production is transformed into juice, but in large producing countries (Brazil and the United States), this percentage reaches 96%, which generates a large amount of waste [63]. These figures are relevant as they demonstrate the abundance of one of the main sources of Limonene. Approximately 50,000 tonnes of R-(+) Limonene are processed each year as a by-product of the global coal industry [64,65].

Limonene is a very versatile compound in terms of applications. It is used as a solvent for resins, paints, the synthesis of other chemical compounds, a dispersing agent for oil and in rubber applications [12]. In medicine it is used as an anti-tumour and anti-inflammatory agent [66-68]. In the industrial sector it is an excellent degreaser, effective in cleaning engines, gears, bearings, metal handles and industrial floors.

Some studies point to the use of Limonene as a green corrosion inhibitor for augers in acidic media. Hammouti et al. [69] studied the action of Limonene in various concentrations in combating corrosion of auger in this environment.

The tests were carried out using mass loss measurements, potentiodynamic polarisation and electrochemical impedance spectroscopy. The results showed that the extract acted as a corrosion inhibitor with an inhibition efficiency of 72%.

Manssouri et al. [70] investigated the inhibitory effect of the essential oil extracted from the fruits of the *Ammodaucus leucotrichus* plant, composed of 12.5% Limonene, on the corrosion of carbon auger in a 1M HCl medium. The maximum efficiency of 85.06% was obtained at a concentration of 3 g/L. In the impedance tests, the efficiency at this concentration was 85.66%. The polarisation results showed that the inhibitor under study acts in a mixed way, reducing both anodic and cathodic inclinations. The authors presented the Langmuir isotherm as the adsorption mechanism, thus indicating the formation of a monolayer of the inhibitor on the metal surface.

Although there has been some research into the application of this extract as a green corrosion inhibitor, there is only a small amount of scientific work on the subject. A wide variety of conditions can still be explored, such as application in different media and especially in different flow conditions, so that the results obtained are ever closer to the reality of industrial application.

4.6 Electrochemical techniques for monitoring corrosion

4.6.1 Open Circuit Potential (OCP)

Corrosion potential is defined as a characteristic potential assumed by the metal when it undergoes the corrosive process in a given solution of low electrical resistivity [71]. Experimental determination is relatively easy and provides extremely important information, both for corrosion protection techniques, such as natural corrosion inhibitors, and for research into corrosion processes [72].

This measurement is also called the open circuit potential (OCP) and the constancy of this parameter guarantees the stability of the system, enabling subsequent experiments such as impedance and potentiodynamic polarisation [71]. Figure 11 illustrates the variation in the OCP of carbon auger in H2SO4 at various concentrations of the extract of red dove's foot (*Bauhinia purpurea*) tested as a corrosion inhibitor [73].

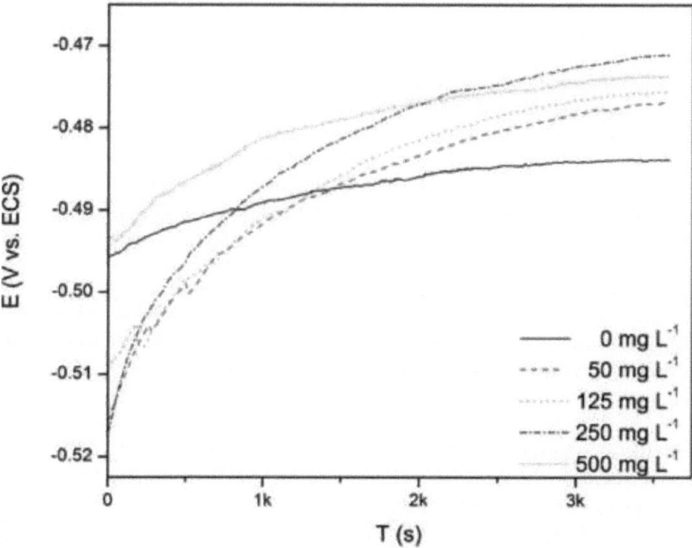

Figura 11: Open circuit potential (OCP) of aug carbon in 1.0 mol.L^{-1} H2SO4 with and without *Bauhinia purpurea* extract at the following concentrations: 50, 125, 250 and 500 mg.L^{-1} at 25°C [73].

It is possible to notice an increase in OCP values with an increase in the concentration of the inhibitor, i.e. the presence of the extract shifts the curves towards more anodic potentials compared to the absence of the extract. According to the authors, this is a partial indication of the extract's effectiveness as a corrosion inhibitor [73].

4.6.2 Electrochemical Impedance Spectroscopy (EIE)

Electrochemical Impedance Spectroscopy is a technique often used to analyse electrochemical processes, through which it is possible to understand the phenomena that occur at the electrode-corrosive medium interface, such as charge transfer, compound adsorption and material transport, among others [74]. In this technique, potentials are applied in frequency ranges that can vary from 10^{-3} to 10^{4} Hz, with the aim of understanding the process in its entirety. The technique does not damage the electrode because the amplitude of the potential wave is small [75].

Figure 12 illustrates the equivalent circuit that simulates the simplest electrochemical process occurring at the electrode-solution interface. This makes it possible, for example, to determine the electrolyte resistance (R_s), the capacitance of the double layer (C_{dl}) and the charge transfer resistance (R_{ct}) of a working electrode (ET).

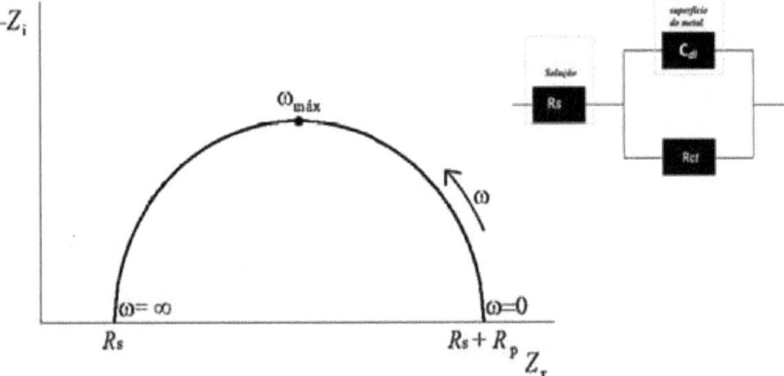

Figura 12: Nyquist diagram for a simple electrochemical system and its representation of the equivalent circuit of the metal/solution interface under corrosion. Adapted from Wolynec [71].

Wolynec [71] presents a systematic physical treatment relating the variables that exist in this technique. In an alternating current circuit, the electric potential E(t) varies with time t and can be represented by Equation 4.11:

$$E(t) = A.\cos\omega t \qquad (4.11)$$

Where A is the amplitude of the electric potential signal and *a>* is the angular frequency expressed in radians per second, given by 2rcf, where f is the oscillation frequency of the signal expressed in Hertz (Hz).

The response of the current I(t) to this potential oscillation is as shown in Equation 4.12:

$$I(t) = B.sen\,(\omega t + \phi) \qquad (4.12)$$

Where B is the amplitude of the current signal and ϕ is the lag of the current in relation to the potential (between the disturbance and the response), also called the phase angle. The relationship between the potential E(t) and the current I(t) can be expressed by Equation 4.13:

$$E(t) = Z.I\,(t) \qquad (4.13)$$

Where Z is the electrode impedance.

The impedance (Z), being a vector, has a modulus |Z| and a phase angle (ɸ). This vector can be represented on a complex plane as a function of its real component (z_r) and its imaginary component (z_i) as described in Equation 4.14:

$$Z = Z_r + jZ_i \qquad (4.14)$$

Experimental results can be analysed using graphical representations such as the Nyquist and Bode diagrams. The Nyquist diagram is formed by plotting the experimental values of the Zr and Zi components of an electrochemical system on a graph of -Zi as a function of Zr (see Figure 12). As this is a simple electrochemical process, this diagram can be represented by a single capacitive arc, resulting in equality between the value of the charge transfer resistance (R_{ct}) and the polarisation resistance (R_p) [71]. However, more complex systems have circuits with different configurations.

The advantages of the *Nyquist* diagram are that it is easy to observe the effects of impedance in electrochemical tests. At high frequencies, the system impedance is given exclusively by the solution resistance (R_s) and at lower frequencies, the impedance is represented by the sum of the solution and polarisation resistances (R_s+R_p) [74].

The Bode diagram shows the variation in absolute impedance |Z| and phase angle (ɸ) as a function of frequency (w), as shown in Figure 13.

The |Z| x log(f) curves in Figure 13 show that, at high frequencies, the ohmic resistance prevails over the system impedance and R_s can be identified by extrapolating the y-axis to the horizontal platform generated by the curve. For small frequency values, Rs+Rp is obtained [75].

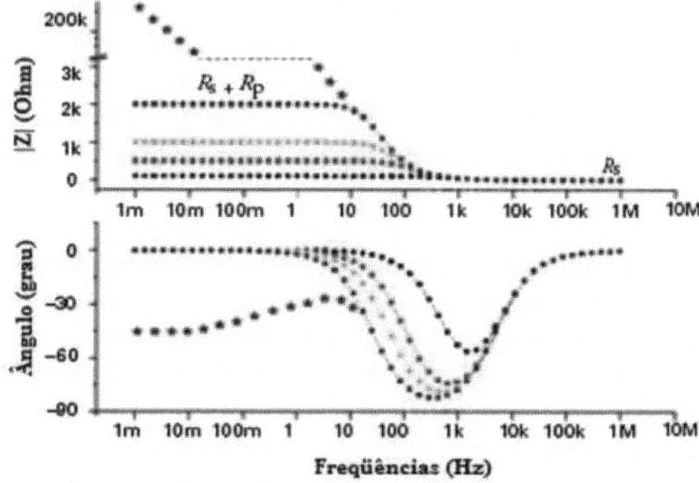

Figura 13: Bode diagram [76].

In the Bode plot relating the phase angle (ɸ) x log(f), it can be seen that at frequencies close to zero, the phase angle is also close to zero, justifying the resistive behaviour of the electrochemical system and that the value of the capacitance of the double layer is obtained at the maximum phase angle (ϕ_{max}).

4.6.3 Potentiodynamic Polarisation Test

In this method, the metal to be analysed is called the working electrode (WE) and the medium under investigation is called the electrolyte whose interaction is to be investigated. In addition to the working electrode, two other electrodes are used: the reference electrode (RE) and the counter electrode (CE).

In the test, the current values are measured as a function of the applied potential (*I=f(E)*), thus generating the polarisation curves of the system under study [15]. The curves are obtained by continuously varying the electrode potential E by applying overvoltages (q_s) in both the anodic and cathodic directions, thus polarising the metal and recording the variation in Ai, which also varies continuously [71].

The most widely used method for analysing polarisation curves is the Tafel extrapolation method, as it is efficient and provides parameters of fundamental importance in research, such as: cathodic coefficients (β_c), anodic coefficients (β_a), as well as corrosion potential (E_{corr}) and corrosion current density (i_{corr}) [15]. The graphical representation of Tafel's law is shown in Figure 14.

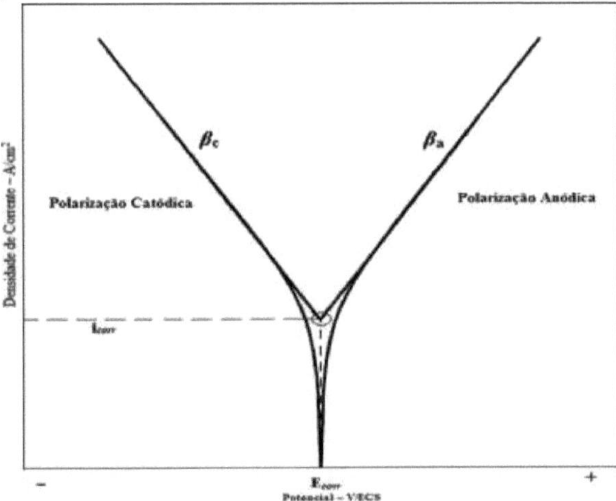

Figura 14: Tafel curves showing how j_{corr}, E_{corr} and the Tafel constants B_a and B_c are determined - Adapted from Gentil [15].

4.7 Infrared Spectroscopy

Fourier Transform Infrared Spectroscopy (FTIR) is one of the most widely used analytical techniques for analysing samples, since it allows investigation of practically any physical state of matter. Using this technique, it is possible to analyse solutions, organic extracts, pastes, powders, films, fibres and surfaces, taking into account the correct choice of sampling technique and preparation route [77].

The frequency range corresponding to the mid-infrared region of the electromagnetic radiation spectrum lies between the wavelengths of 2.5 *pm* and 25 *pm* or between 4000 and 400 cm^{-1} as illustrated in Figure 15. This is the range most useful for identifying and analysing materials [78].

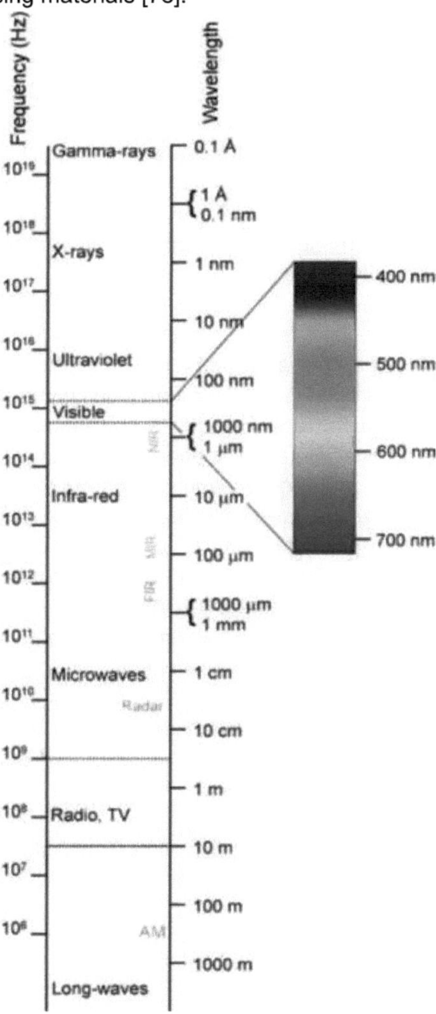

Figura 15: Electromagnetic radiation spectrum [78].

4.7.1 The infrared absorption process

The technique is based on the vibrational behaviour of atoms in a molecule when they are disturbed by infrared radiation. The spectrum generated is a result of the infrared radiation passing through the sample and the fraction of the incident radiation absorbed at each frequency is identified and interpreted by energy or wavelength.

This absorption process excites the molecules, which are then elevated to higher energy states. As this is a quantised process, only specific frequencies are absorbed by the sample, with the corresponding energy variation being between 8 and 40

kJ/mol^2 [78].

Although the absorption process in the infrared is quantised, the final spectrum is made up of numerous overlapping lines whose association constitutes bands directly related to the vibrational and rotational movements of the molecules. Factors such as the relative mass of the atoms, the bond strength constant, the structure and geometry of the compound's structure all have a direct effect on the variation in the frequency or wavelength of absorption of radiation [79].

Infrared spectra are presented in terms of wavelength (pm) or wavenumber (cm^{-1}) on the abscissa axis and band intensities in terms of percentage transmittance (%T) or absorbance (A) on the ordinate axis.

4.7.2 Molecular vibration modes

Molecular vibrations are classified into axial and angular deformations. When there is a rhythmic movement along the axis that connects the molecules, causing an alternating increase and decrease in the interatomic distances, then we have an axial deformation vibration. When these deformations correspond to rhythmic variations involving changes in the bond angles in relation to a set of arbitrary coordinates of the molecule, we have an angular deformation vibration [79].

The total number of degrees of freedom of each atom is equal to the number of degrees of freedom of its respective molecule. In other words, each orientation in the Cartesian plane corresponds to one degree of freedom. In this way, a molecule with *n* atoms will have *3n* degrees of freedom. In molecules with non-linear geometry, three of the degrees of freedom describe translation and three describe rotation. The remaining 3n - 6 degrees of freedom are associated with the fundamental vibrations of the molecule [80].

In general, molecular bonds have six vibrational modes, as shown in Figure 16.

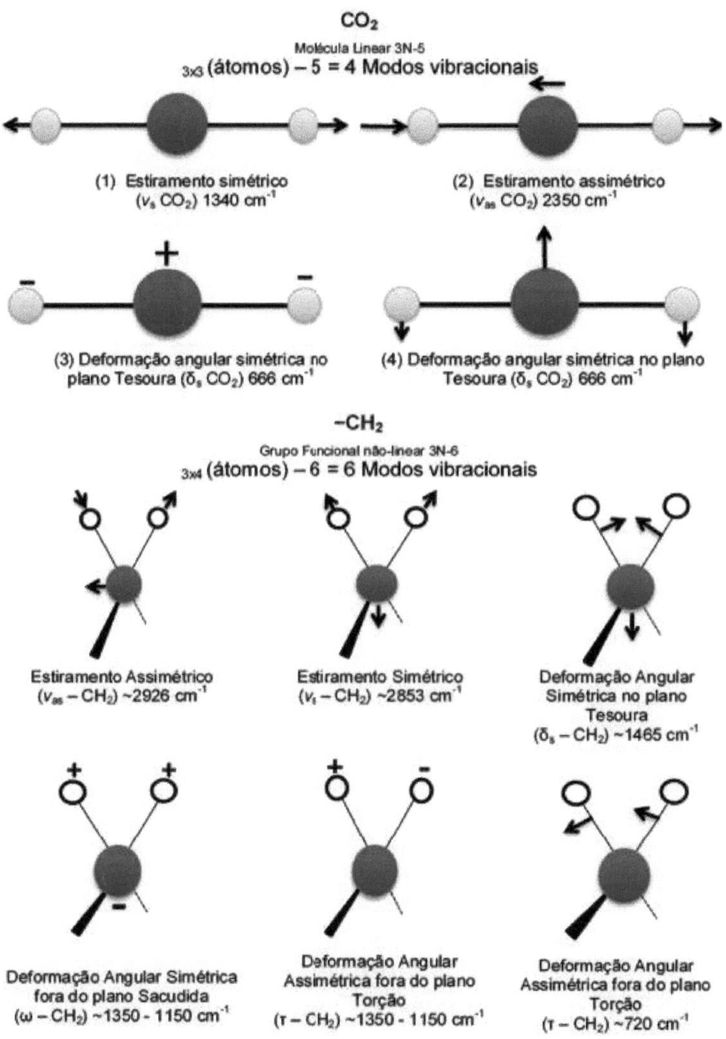

Figure 16: Vibrational modes for the CO2 molecule and the CH2 functional group. The + and - signs represent movements perpendicular to the plane of the page. Adapted from Silvertein [79].

4.7.3 Instrumentation

Spectrophotometers, depending on their operating mechanism, can be classified as dispersed beam or Fourier transform spectrophotometers.

In dispersion spectrophotometers, the energy source emits a beam of radiation that passes through the sample, making it possible to obtain spectra. Although the operating mechanism is similar to the Fourier Transform, it has several advantages over dispersion instruments.

The fact that they don't have monochromators results in extremely high spectral

resolutions (< 0.001 cm^{-1}), as the entire radiation band passes through the sample in a short time. After numerous combined scans, it is possible to reduce the noise generated, thus obtaining high quality spectra with a small amount of sample [80].
Figure 17 illustrates how an FTIR spectrometer works. The fundamental principle of the equipment is the emission of infrared radiation by the radiation source. This radiation covers all the wavelengths of interest and is divided into two beams that fall on a fixed mirror (fixed distance) and a moving mirror (variable distance).

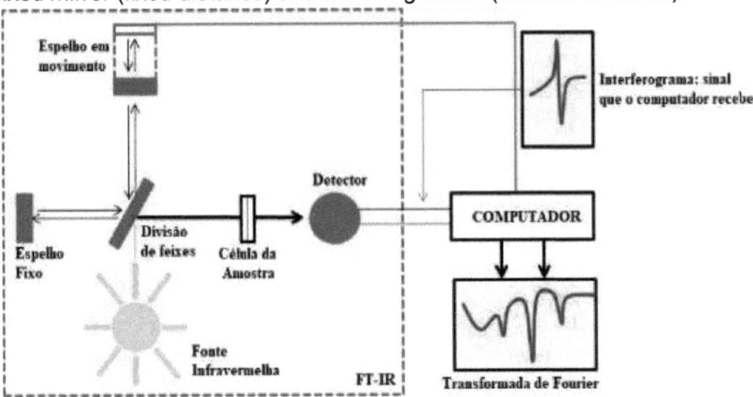

Figura 16: Schematic of the optical system of an FTIR spectrometer. Adapted from Martins [78].

The reflection of the beams in both mirrors generates a sequence of constructive and destructive interferences and the variations in radiation intensity are sensitively recorded by the detector, also known as an interferogram. The interferogram provides information in the time domain that is complex to encode and impossible to analyse in practice. It is therefore not possible to obtain any information about the sample being analysed from the information in this component.
Thus, the application of the Fourier Transform converts the time domain of the interferogram into the frequency domain, making investigative analyses possible. Variations in the position of the movable mirror attached to a track also vary the distance travelled by the beam. The Fourier Transform is applied to the result of each successive change in the position of the mirror, thus generating a complete infrared spectrum. In general terms, the radiation that passes through the sample hits it with a specific band of energy that gives rise to the complete spectrum of the sample [80].

5 MATERIALS AND METHODS

The following sections will present the experimental stages, materials and equipment used to obtain the results that will be described, as well as the analysis methods used in this work. In accordance with the objectives, the methodology of this work is based on the following stages, as shown in Figure 18:

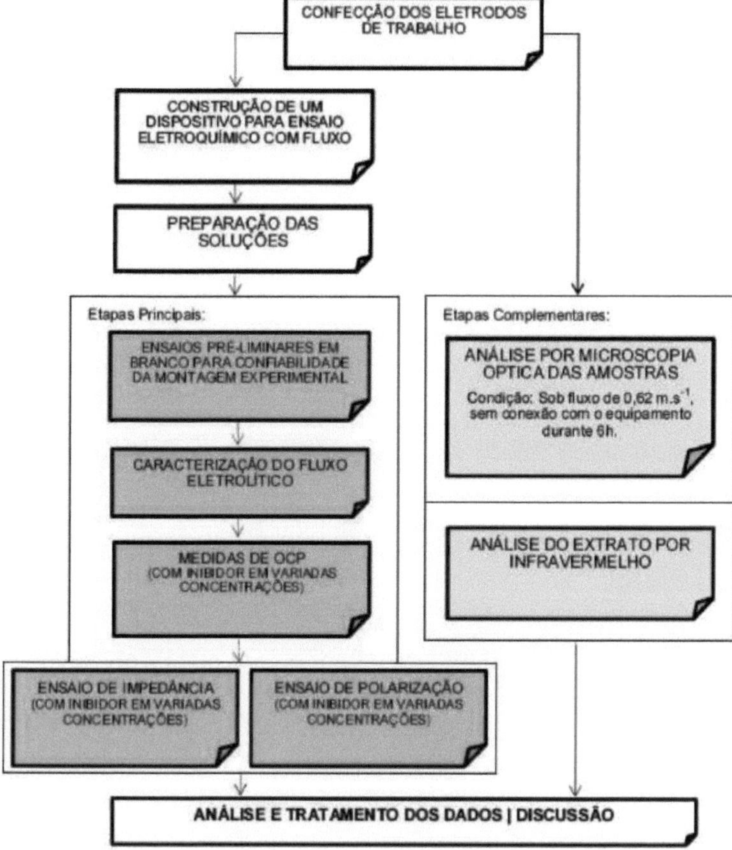

Figure 18: Flowchart of the research methodology.

5.1 Methodology

This work is divided into six experimental parts. Part I consists of characterising the electrolytic flow. Although the main objective of this study is to evaluate the action of the Limonene extract as a green corrosion inhibitor for ABNT 1020 carbon steel in a turbulent 3% NaCl medium, reliability tests were first carried out on the device built to carry out the electrochemical tests (part II).

Preliminary tests were then carried out with various extracts/oils in order to choose the right extract to work with (part III). Part IV is dedicated to a specific study of the Limonene extract due to the factors highlighted in the previous stage.

The speed chosen for the tests in the presence of Limonene extract in various concentrations was 0.62 m.s^{-1}, equivalent to 2.91 L/min or 6583.58 Reynolds. This speed was chosen because it was highly aggressive in the functionality tests, as demonstrated by the low resistance to load transfer, high corrosion rate and high current density.

Part V presents microscopic analyses of the samples from extended tests and Part VI consists of the characterisation of the extract by Fourier Transform Infrared Spectroscopy (FTIR).

5.2 Equipment

The equipment needed to carry out this project is listed in Table 3 below:

Table 3: Specifications of the equipment used in the research.

Equipment	Name/Model	Property
	BIOPRECISA analytical balance - FA2104N (210g x 0.0001g)	IFMA/DMM/ LABMCD
	FISATON Mechanical Shaker - Model 710 (170 to 2000 rpm)	IFMA/DMM/ LABMCD
	Potentiostat with FRA32M Module AUTOLAB - PGSTAT 302N	IFMA/DMM/ LABMCD
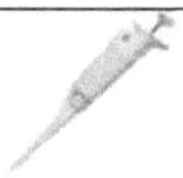	Micropipette KASVI Basic 154935 (Volume: 100 - 1000 μL)	IFMA/DMM/LABMCD
	IFTA 5020 Asymmetrical Digital Power Supply (T: 0 to 30V and C: 0 to 20A)	IFMA/DMM/LABMCD
	Ag/AgCl Reference Electrode (Saturated) RE1-CP 013503 (E = 0.197V at 25°C)	IFMA/DMM/LABMCD
	FTIR spectrophotometer SHIMADZU IRPrestige-21 (S/N > 40000:1 and Res. of 4 cm-1)	UFMA/Analytical Centre
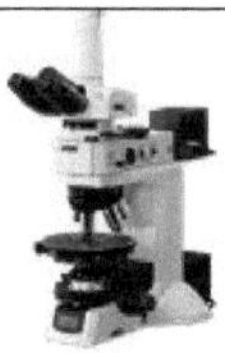	NIKON Eclipse LV100D-U microscope	IFMA/DMM/LABMCD

5.3 Making the working electrodes (ET)

At this stage, 10 mm diameter ABNT 1020 bars were machined to make the working electrodes (TE), which have a diameter of 5 mm x 13 mm in length. They were then coated with hard rubber for a perfect fit in the electrochemical cell. The working electrode model used in the tests is illustrated in Figure 19.

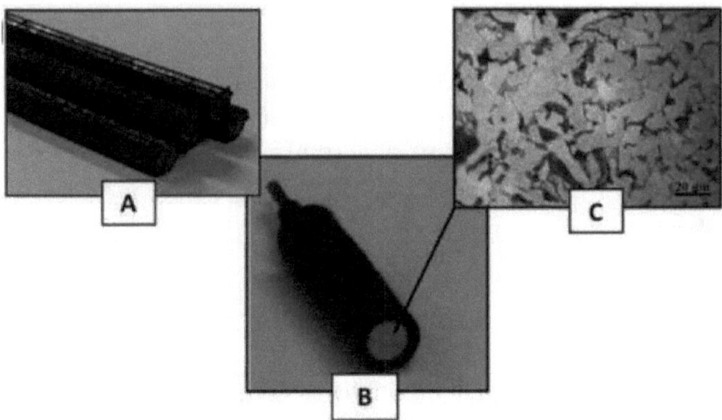

Figura 19: Working electrode model used in the experiments. A) Bars in the as-received condition, B) ET and C) Optical microscopy of ABNT 1020.

Table 4 shows the composition of ABNT 1020 as received:

Table 4: Composition of ABNT 1020.

Material	C%	Mn%	Si%	P% (Max.)	S% (Max.)
ABNT 1020	0,18	0,30	0,05	0,04	0,05

After the electrodes had been machined and embedded, they were sanded with water-based siKcium carbide sandpaper ranging from 100 to 1200 mesh. Considering the small surface area exposed (0.2 cm^2) and the difficulty in sanding it, a small support was used to sand the electrodes in order to guarantee the flatness of the surface. After sanding, the electrodes were washed with neutral soap in running water to remove any grease and then dried with a hot air jet.

The working electrode is securely connected to the potentiostat by soldering a copper wire into a small 3 mm hole at one end of the electrode, as shown in Figure 20.

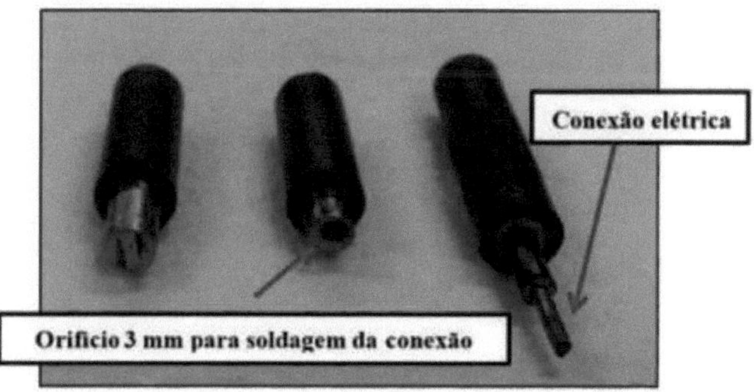

Figura 20: Stages in the preparation of working electrodes.
^ External diameter of the electrode (d) = 5.00 mm;
+ External diameter of the electrode with side coating (d') = 8.00 mm;
+ Electrode length (c) = 13.00 mm; and
^ Surface area exposed to electroKtic flow (A) == 0.2 cm .[2]

5.4 Electrochemical system for flow testing

At this stage, a 150 mm x 150 mm x 170 mm glass reservoir was made to contain the solution (3% NaCl or 3% NaCl + Yppm Limonene) which will circulate through 10 mm diameter acrylic tubing via a 12V DC mini-pump with a maximum rated current of 350 mA, 5 W and a maximum flow rate of 240 L/h which will drive the flow of the solution.
The electrochemical cell is made up of three electrodes: the working electrode (ET), the reference electrode (ER) and the counter electrode (CE). The ER will be made of Ag/AgCl (7 mm) and the CE (6 mm) will be made of graphite to close the circuit. The electrochemical cell was made from Nylon material at the facilities of the Department of Mechanics and Materials - DMM/IFMA. To ensure that the electrodes fit perfectly into the electrochemical cell and to avoid possible leaks, sealing rings were used around the electrodes, thus guaranteeing a process without loss of solution.
The electrodes were arranged so that their faces were in contact with the flow of the solution, as illustrated in Figure 21. This positioning is based on the work presented by Leon et al [59]. The 3D representation of the electrochemical cell and its dimensions is shown in Figure 22.

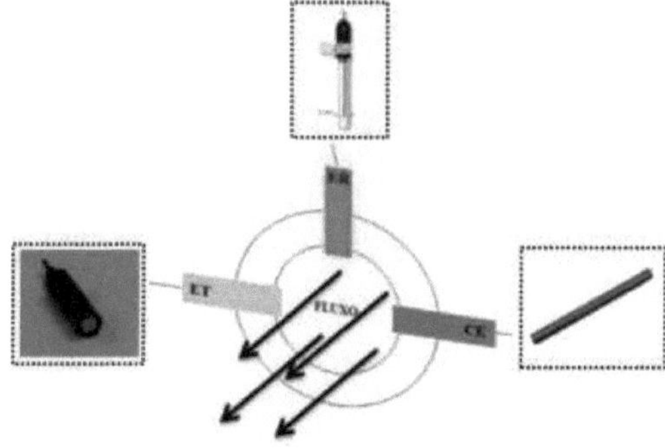

Figura 21: Position of the electrodes in the electrochemical cell. ET - Working electrode, ER - Ag/Ag Cl reference electrode, CE - Graphite counter electrode.

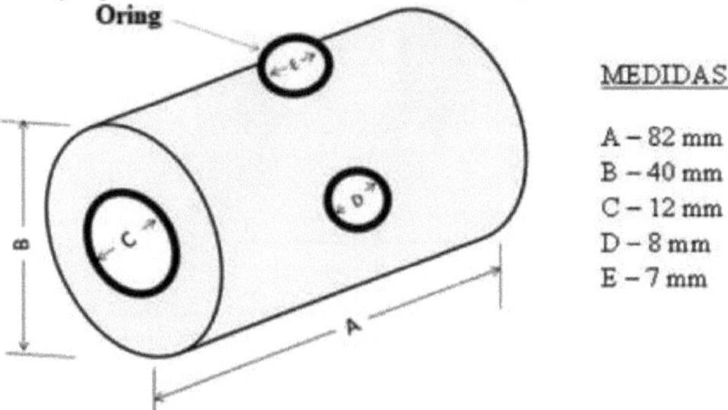

Figura 22: 3D representation of the electrochemical cell and its dimensions.

Flow is a fundamentally important parameter that needs to be measured. For this purpose, a YF-S201 model %" flow sensor was used, with an operating voltage ranging from 5 to 24 V and a maximum current of 15 mA (5V), which can measure a flow of up to 30 L/min. It is powered by the Hall effect, which enables flow to be measured in real time. This sensor will be connected to a small board via an arduno system located at the output of the electrochemical cell for better control, thus avoiding loss of charge from other connections.

The arduno UNO system integrates the Flow Reader Unit (ULF), whose aim is to identify the velocity of the electrolyte and characterise the flow as laminar (EL), transitional (END - Flow Not Defined) or turbulent (ET) using the Reynolds number. This information will be accessible during the experiment via a 16x2 LCD display based on the equations of fluid dynamics. The power supply of this unit is compatible with the output voltage of the USB port of the computer used to obtain the curves.

The mini-pump chosen is completely waterproof and suitable for working continuously with fluids such as water, oil, petrol, acid and alkaline solutions. It is powered by a power supply model ITFA/5020 belonging to the LABMCD - Multiuser Corrosion and Wear Laboratory of the Department of Mechanics and Materials at IFMA.

To avoid chemical contamination of the solution by dissolution of any metallic material from the prototype during the experiment, care was taken to use non-metallic materials in the pump components that maintain contact with the electrolyte, as well as the connections used, all of which can be adjusted and easily removed for cleaning after the tests. The complete experimental setup is shown in Figure 23. The experimental set-up is modelled on the work by Leon et al. [59], plus a Flow Reader Unit-ULF.

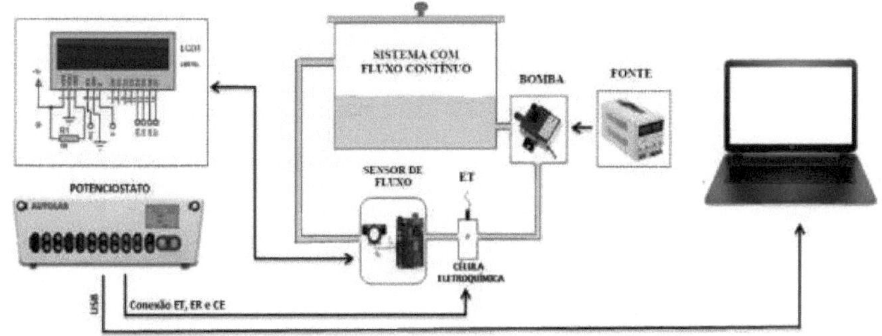

Figura 23: Diagram showing the positioning of the experiment components.
Before the physical construction, the electronic model of the Flow Reader Unit was designed and simulated in the Proteus software® 2.1 (as shown in Figure 24) with the mathematical formulations copied into a programming language stored in the Arduino UNO device that allows the flow rate to be read from the PWM (*Pulse Width Modulation*) pulses emitted by the flow sensor. The main elements of the experimental set-up are shown in Figure 25.

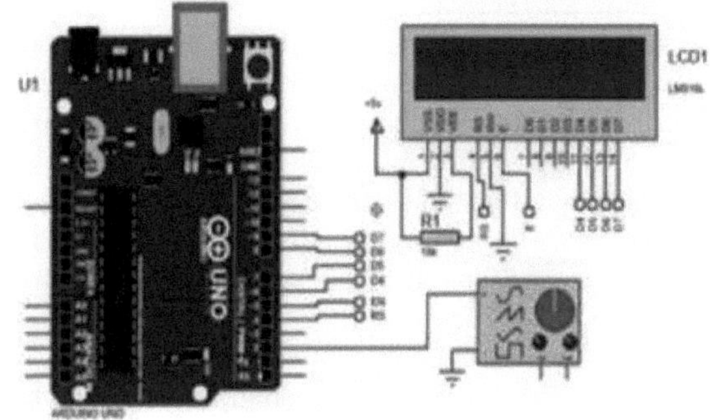

Figura 24: Schematic of the Flow Reader Unit simulated in Proteus Software® 2.1.

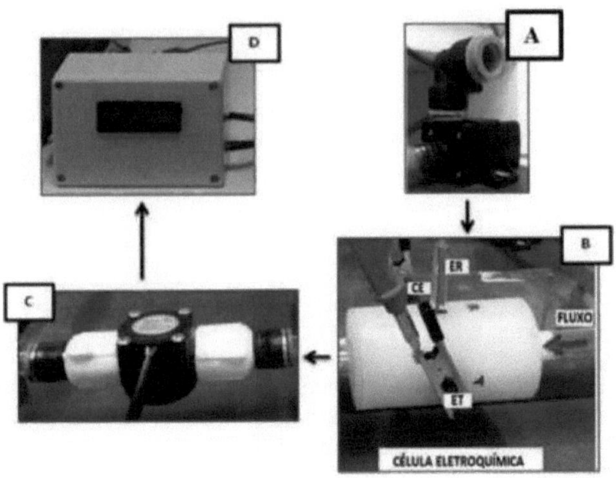

Figura 25: Highlighting the elements of the experimental bench. A) Mini-pump, B) Electrochemical cell, C) Flow sensor and D) Flow reader unit - ULF.

5.5 Obtaining Limonene Extract

The Limonene extract tested as a natural corrosion inhibitor was purchased from Destilaria Bauru, a company that produces, imports and exports various types of natural extracts and vegetable oils located at Rua Rio Grande do Norte, 80, Higienopolis in Catanduva -SP - Brazil.

The volume of extract used in this research belongs to batch No. DBISU-DLMNN2133081, reference PRO194, catalogued under CAS 5989-27-5, under the description D'Limoneno de Laranja, scientific name *Limonene*, manufactured on 11 July 2017 and valid until 11 July 2019.

Steam distillation was the method used to extract limonene, as this is the most widely used procedure for obtaining essential oils worldwide. Using this method, it is possible to obtain extracts from a wide variety of plant species, such as seeds, bark, leaves, roots and grasses.

The distiller, also called an extractor, is the equipment used in this process. In a nutshell, the vapour passes through the tissues of the plant material, extracting the oil from inside the glands, which in turn vaporises as a result of the thermal shock towards the condenser. In the condenser, the mixture of oil and hydrolate (by-product) cools and returns to the liquid phase. Immediately after this stage, the process of separating the phases begins using the separator vessel.

After the entire distillation process, the extract was evaluated in some respects as shown in Table 5 and 6.

Table 5: Assessment of colour, odour and appearance. Information taken from the extract's Certificate of Analysis, as per ANNEX I.

Item	Description
Colour	Colourless/Transparent
Odour	Orange Characteristic

Aspect	L^quid

Table 6: Assessment of aldehyde content, density and refractive index. Information taken from the extract's Certificate of Analysis, as per ANNEX I.

Analysis item	Specification	Results
Aldehyde content (%):	0,37 - 1,50	0,80
Density:	0,838 - 0,840	0,84
Refractive index:	1,470 - 1,475	1,471

5.6 Preparation of solutions

For all the tests, 3% NaCl solutions (pH ~ 6.8) were prepared from distilled water. Each test required 650 ml of 3% NaCl solution in order for the system to work perfectly, ensuring that the electrodes were always in contact with the solution.

The extracts were then dosed using a KASVI Basic 154935 Micropipette (100 - 1000 pL), a single-channel mechanical micropipette with adjustable high-precision settings, and then mixed with 650 ml of 3% NaCl solution using a Mechanical Shaker at 1000 rpm for 30 minutes, as shown in Figure 26.

Figure 26: Beaker containing 250 ppm of Limonene extract being shaken at 1000 rpm by the Mechanical Shaker of the IFMA Multiuser Corrosion and Wear Laboratory.

5.7 Electrochemical tests

Potentiometry (OCP) measurements were taken, followed by electrochemical impedance measurements. New OCP measurements were carried out to ensure that the system was stabilised, followed by Potentiodynamic Polarisation measurements.

In the impedance tests, the frequency applied varied from 105 to 1 Hz, with a disturbance amplitude of 10 mV and 10 points/decade. The equivalent circuit was generated in the software that is ported to the equipment (Nova 1.1®) and then the values obtained in the experiments were transferred to the Origin software® to be treated statistically and analysed.

For the Potentiodynamic Polarisation tests under the influence of flow, the scan rate applied was 1 $mV.s^{-1}$ with anodic and cathodic overvoltages of +200 and -200mV, respectively, in relation to the open circuit potential (OCP). The Tafel Extrapolation method used to obtain the parameters follows the methodology presented by Poorqasemi [81]. All the experiments were carried out in triplicate for greater reliability of the results and at room temperature (24°C ± 1°C).

At the end of each electrochemical test, all the connections and elements of the bench (sensor, mini-pump and electrochemical cell) were removed, cleaned with neutral soap and dried with a hot air jet, in order to guarantee the accuracy of the extract concentrations for participation in the tests.

5.8 Obtaining OCP, FRA and Polarisation curves under the influence of turbulent flow

The flow analysis device will be connected to the potentiostat and linked to the computer, which will make it possible to produce the curves using the NOVA software version 1.1® . The work is based on the ASTM G59- 97 standard [82], which deals with the positioning of the electrodes and the connections between the computer, electrochemical cell and potentiostat. The tests will be carried out at room temperature.

Figure 27 illustrates the PGSTAT 302N Potentiostat/Galvanostat belonging to the LABMCD - Multiuser Corrosion and Wear Laboratory of the Department of Mechanics and Materials at IFMA/Campus Monte Castelo.

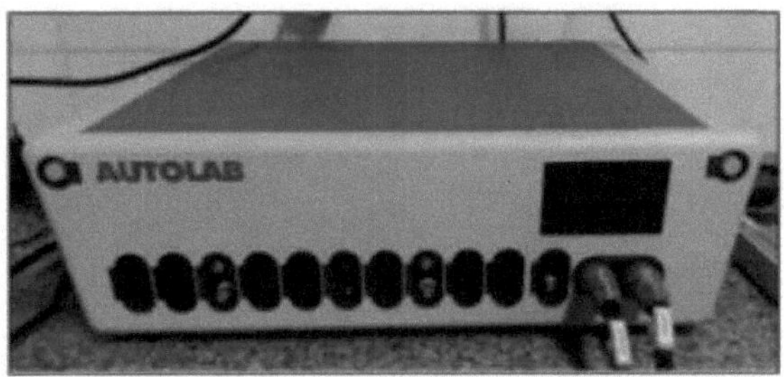

Figura 27: PGSTAT 302N potentiostat/galvanostat used to obtain OCP, Potentiodynamic Polarisation and Electrochemical Impedance curves.

With the electrodes duly positioned, the experiment begins by first determining the open circuit potential (OCP) of the metal in relation to the electrolyte under study to ensure stabilisation of the electrochemical system. Once the OCP curves have been drawn, we move on to the Electrochemical Impedance test and then the Potentiodynamic Polarisation test.

The programming script for surveying the curves follows the standard procedure, since the Impedance tests must precede the Polarisation tests. This programmable sequence is justified by the fact that impedance tests do not damage the sample, unlike polarisation tests.

27.9 Standards and formulations for processing electrochemical data

Through the impedance tests it will be possible, by means of the Equivalent Electrical Circuit (EEC), to extract the charge transfer resistances in the presence ($R_{ct,o}$) and in the absence of the Limonene extract (R_{ct}). The inhibition efficiency (EI) and the degree of inhibitor coverage (0) in this environment can be calculated using Equations 5.1 and 5.2 respectively [42] [44,45]:

$$EI = \frac{R_{ct,o} - R_{ct}}{R_{ct,o}} x\,100 \quad (5.1)$$

$$\theta = EI\ x\ 100 \quad (5.2)$$

Through polarisation tests, Tafel slopes will be applied to determine the cathodic ($_{BC}$) and anodic ($_{Pa}$) coefficients. It will also be possible to extract the corrosion currents in the presence ($_{ic}$) and absence of the inhibitor ($_{isi}$). The efficiency of the inhibitor (n%) in this environment can be calculated from Equation 5.3:

$$\eta\% = \frac{i_{si} - i_{ci}}{i_{si}} x\,100 \quad (5.3)$$

The work is based on the ASTM SI 10 standard [83], which regulates the conventional units for corrosion tests, and the work by Hammouti et al. [69], which cites Equation 5.3 based on the ASTM G184 - 06 standard [84] for calculating the efficiency of corrosion inhibitors.

27.10 Optical Microscopy

After the electrochemical tests, samples were prepared, sanded, polished and subjected to prolonged tests without connection to the potentiostat for a period of 6 hours in order to obtain visual proof of their preservation in the presence of the Limonene extract.

The samples were exposed to turbulent flow under the same conditions used in the electrochemical tests. The microscopes were obtained using a Nikon Eclipse LV100D-U optical microscope associated with an image acquisition system belonging to the Department of Mechanics and Materials (DMM) of the Federal Institute of Education, Science and Technology (IFMA).

27.11 Fourier Transform Infrared Spectroscopy - FTIR

To carry out this experimental stage, an FT-IR spectrophotometer with an ATR/MIR cell was used. The analyses were carried out in transmittance using KBr pellets, operating with 45 scans with a resolution of 4 cm^{-1} and in the wavenumber range between 400 cm^{-1} and 4000 cm^{-1} . In general, the Limonene sample is inserted into the holder, which in turn is positioned in the equipment's sample holder. All the parameters for acquiring the spectrum are entered into the software that is ported to the equipment.

The records were initially obtained in transmittance mode and the characteristic groups of each band were identified. Figure 28 shows a Shimadzu IR-Prestige Spectrophotometer belonging to the Analytical Centre of the Postgraduate Chemistry Programme - PPGQUIM at the Federal University of Maranhao - UFMA.

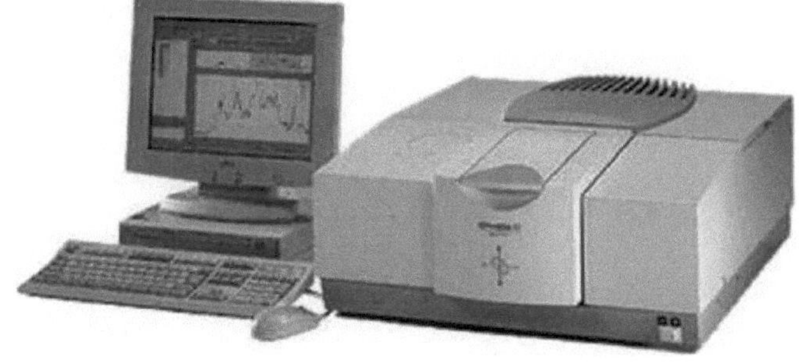

Figura 28: Shimadzu IR-Prestige spectrophotometer used for FT-IR analysis. The equipment belongs to the Analytical Plant of the Postgraduate Programme in Chemistry - UFMA.

The use of this technique is based on ASTM E1252 [85]. This standardises and describes the correct procedures for qualitative analysis of organic and inorganic samples using infrared spectroscopy. It provides details on sample preparation, obtaining spectra and guidelines for studying and interpreting spectra.

6 RESULTS AND DISCUSSION

6.1 Part I: Standard characterisation of electrolytic flow

The kinematic speed readout on the device's LCD display (ULF) is the result of PWM pulses being emitted by means of a kite-shaped valve that works in conjunction with a magnet attached to its structure. Each pulse measures approximately 2.25 mm and the emission of 450 pulses is equivalent to 1L. The frequency of rotation of the flow sensor's vanes is associated with the flow rate Q of the electrolyte at that moment. For a better understanding and characterisation of the electrolytic flow, here is a sequence of the mathematical formulations that make up the programming used to obtain the results shown on the display:

1a Relationship between the frequency of the flow sensor and the flow rate Q (L/min):

$$Frequência_{sensor}(Hz) = 7Q\ (L/min) \quad (6.1)$$

2a Conversion of the flow rate Q (L/min) into Q'(m3/s):

$$Q' = Q.1{,}666.10^{-5} \quad (6.2)$$

3a Relationship between electrolyte velocity as a function of flow rate (Q) → V(Q):

$$V = \frac{Q'}{A} \quad (6.3)$$

Where A is the area in m2 of the pipe used in the device to circulate the equivalent electrolyte $\frac{\pi D^2}{4} = \frac{(\pi.0{,}01^2)}{4} = 7{,}85.10^{-5}\ m^2$. **We therefore have**

$$V = \frac{Q'}{A} \therefore V = \frac{Q.1{,}666.10^{-5}}{7{,}85.10^{-5}} \therefore V = 0{,}2122Q \quad (6.4)$$

4a Calculation of the Reynolds number (Re) as a function of V, D and v:

$$Re = \frac{V.D}{v_{NaCl\ 3\%}} \quad (6.5)$$

Since *v* is the kinematic viscosity of 3% NaCl equivalent to 0.93765.10-6 m2/s, we have:

$$Re = \frac{0{,}01V}{0{,}93765.10^{-6}} \quad (6.6)$$

5a Classification of flow according to Reynolds number [54].

- For Reynolds less than 2000, laminar flow (LF);
- For Reynolds numbers between 2000 and 2400, transient flow (ETR);
- For Reynolds numbers greater than 2400, turbulent flow (ET);

Figure 29 shows the dependence of the flow rate Q (L/min) on the frequency of rotation of the flow sensor vanes. In the case studied, with information such as the pipe diameter (D) and the kinematic viscosity of 3% NaCl ($v_{NaC\ i\ 3\%}$), it is possible to determine the maximum flow rate (Qmin.) for obtaining turbulent flow using Equation 6.5.

$$Re = \frac{V.D}{\nu_{NaCl\ 3\%}} \tag{6.5}$$

$$\overset{CALC.}{\Longrightarrow} 2400 = \frac{0{,}01V}{0{,}93765.10^{-6}} \therefore V_{min.} = 0{,}225 \frac{m}{s}\ ou\ Q_{min.} = 1{,}06\ L/min$$

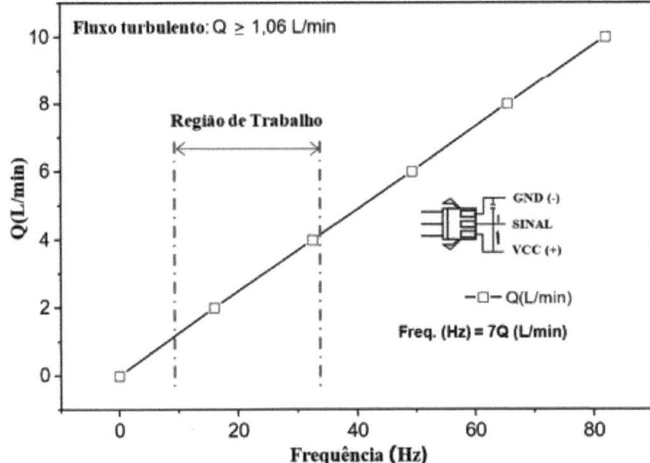

Figura 29: Graphical demonstration of the region of the system with turbulent flow using the flow sensor.

Table 7 describes the working conditions for each flow rate tested in this work.

Table 7: Flow breakdown for each flow rate studied.

Q (L/min)	V (m.s)$^{-1}$	Re	Classification
0,94 ± 0,047	0,20 ± 0,007	2180,34 ± 027,30	Transient Flow
1,93 ± 0,096	0,41 ± 0,018	4372,63 ± 191,97	Turbulent Flow
2,91 ± 0,145	0,62 ± 0,027	6583,58 ± 316,64	Turbulent Flow
3,97 ± 0,198	0,84 ± 0,042	8958,56 ± 447,93	Turbulent Flow

This means that when the pump is fed in such a way as to work with a flow rate greater than 1.06 L/min, the flow regime in question is turbulent, i.e. under these conditions the fluid particles travel in a random direction, since the velocity has transverse components to the general movement of the whole [53,54]. Finally, Figure 30 illustrates the information accessible on the Flow Reader Unit.

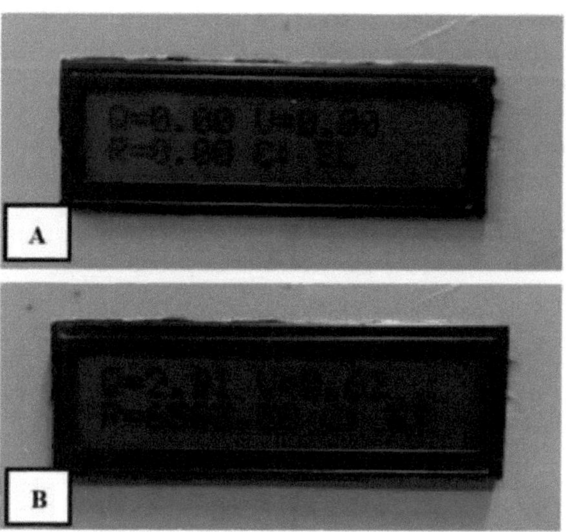

Figura 30: Parameters obtained from the Flow Reader Unit - ULF. A) Initial reading before the electrochemical tests and B) Reading during the electrochemical test of 2.91L/min.

6.2 Part II: Blank tests at different speeds

All the results obtained in this and the other experimental stages involving electrochemical tests were carried out using the experimental bench illustrated in Figure 31. The aim of this stage is to assess the sensitivity of the system to variations in speed, since knowing how it responds to the various speeds tested will make it possible to choose the appropriate speed for a new battery of tests involving the addition of Limonene extract (part IV).

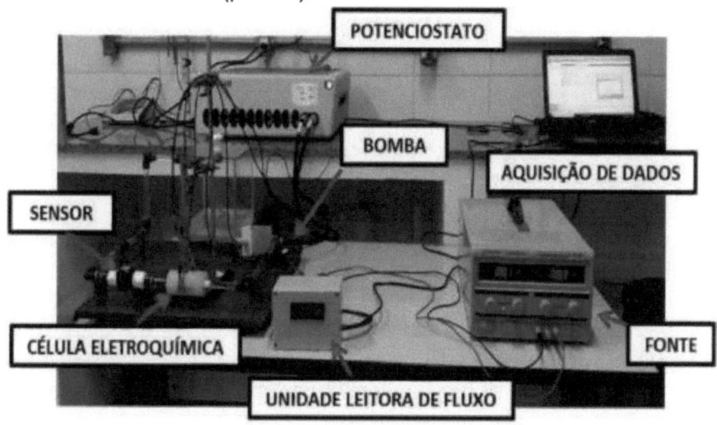

Figura 31: Experimental bench built for electrochemical tests with electrolytic flow control - Multiuser Corrosion and Wear Laboratory - LABMCD - IFMA.

6.2.1 Electrochemical Impedance Measurements

Figure 32 shows the electrochemical spectra obtained at various speeds for the 3% NaCl electrolyte. A specific behaviour representative of the experimental curves can be seen for the different speeds investigated. Although the tests were carried out in the same frequency range (105 to 1 Hz), the influence of the electrolyte flow on the system's real (Z') and imaginary (Z") impedance values can be clearly seen.

In qualitative terms, the arc formed by the experimental points in the absence of flow is larger than the others in the presence of the various flows, which is an indication of the high resistance to charge transfer in a stationary condition.

However, discussions in quantitative terms require the application of an equivalent electronic circuit that better simulates this system and presents a lower standard deviation (/).

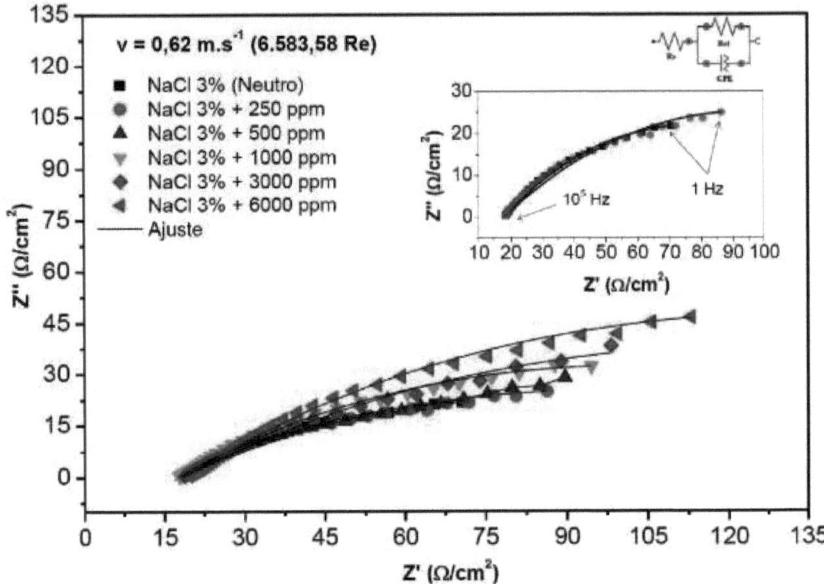

Figura 32: Nyquist spectrum for tests in the absence and presence of flow at different speeds and representation of the equivalent electrical circuit R - [R|Q] used in the simulations.

The Nyquist spectra in Figure 32 were adjusted to an Rs- [Rct|Q] circuit, since this circuit represents the existing electrochemical interactions based on previous studies [71]. The circuit used determines the values of the resistance of the solution (Rs), the resistance to charge transfer (Rct) and the capacitance of the electrical double layer (Cdl) accompanied by the index *n* using the CPE element. The square of the standard deviation (x2) represents the fit between the experimental results and the simulations produced.

The units of the parameters obtained are based on the ASTM SI 10 standard [83], which regulates conventional units for corrosion tests. With this information, Table 8 shows more detailed results after applying the circuit to the electrochemical spectra.

Table 8: Electrochemical parameters obtained through the equivalent electronic circuit at different speeds.

Speed (m.s)$^{-1}$	Lol (Q)	Rct (Q/cm)^{2}	CPE Cdl (mF/cm)^{2}	CPE *n*	X^2
No flow	42,3	330,19	0,131	0,6	0,0178
0,20	18,9	310	1,871	0,49	0,0116
0,41	18,4	249,89	0,951	0,56	0,0228
0,62	18,5	87,75	0,971	0,53	0,0104
0,84	18,2	76,46	0,743	0,55	0,0096

Note: *R_s - resistance of the solution, Rct - resistance to charge transfer, Cdl - capacitance of the electrical double layer, n - index measuring the ideality of the CPE element and x^2 the square of the standard deviation.*

It can be seen that the test carried out in the absence of flow has a maximum load transfer resistance of 330.19 Q.cm^{-2} , since this condition is less aggressive to the surface of the material due to the absence of diffusive effects caused by the intensity of the flow.

Outside the stationary state, it can be seen that the Rct values decrease as the electrolyte velocity increases. This is related to the greater diffusion of the species involved in this dynamics, which in turn reach the metal surface with greater intensity. The test carried out at 0.20 m.s^{-1} (2180.34 Reynolds) does not guarantee turbulent flow, but rather transitional flow. In this condition, a Taylor vortex forms in the space between the electrodes and operation in this flow regime is not advisable for electrochemical studies [52].

A resistance of 249.89 Q.cm^{-2} was recorded in the test carried out at a flow rate of 0.41 m.s^{-1} and this value decreased for the tests carried out at higher speeds, reaching a minimum resistance of 76.46 Q.cm^{-2} when carried out at 0.84 m.s^{-1} at a concentration of 3% NaCl. In this speed variation, there is a change in the capacitance values of the double layer, a fact attributed to a change in the structure of the electrical double layer, as well as a change in the dielectric constant of the material [42]. The capacitance values can be calculated using Equation 6.7 [50]:

$$C_{dl} = \left(Y_0 \frac{R_{ct}}{R_{ct}^{n}}\right)^{1/n} \quad (6.7)$$

Where Yo represents the electrical conductance (S), Rct the resistance to charge transfer (Q.cm-2) and n represents the index that measures the ideality of the CPE element, which can vary from 0 to 1.

The *n-index* registered 0.6 in the absence of flow and 0.49 in the transient flow tests. However, for the tests in the presence of turbulent flow, this index varied between 0.53 and 0.56. Although the experimental frequency range is the same (10^5 Hz - 1 Hz), it should be noted that the speed and flow conditions also influence the measurement of this parameter. It is worth noting that the test carried out under transition conditions was the one with the lowest *n-index* (0.49), which may be related to the formation of Taylor's vortex, which makes subsequent approaches and analyses of this data unsuitable for this research [52].

6.2.2 Potentiodynamic Polarisation Measurements

The experimental curves obtained by potentiodynamic polarisation for the various speeds are shown in Figure 33. It is noteworthy that the presence of flow shifts the curves to the right side of the graph, thus indicating an acceleration in the corrosion process.

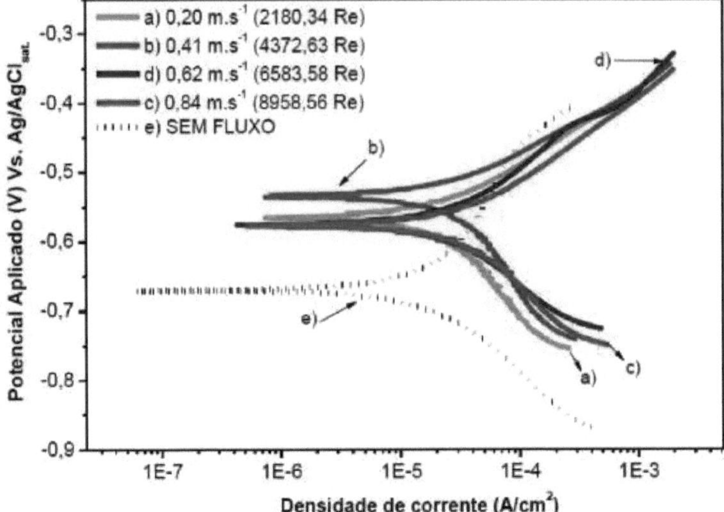

Figure 33: Potentiodynamic polarisation curves for different speeds of 3% NaCl.

The result of applying the Tafel Extrapolation Method to the experimental curves is shown in Table 9 and provides a better understanding of these curves.

Table 9: Electrochemical parameters obtained in the polarisation tests for different speeds of 3% NaCl.

Speed (m/s)	Ecorr (mV)	Jcorr (pA/cm)^2	eanodic (mV/dec)	TC (mm/year)	RP (Q)
No flow	-670,55	112,93	172,49	1,31	2140,03
0,20	-524,86	144,78	270,14	1,68	1279,90
0,41	-532,79	178,42	310,13	2,07	992,94
0,62	-575,28	217,69	213,34	2,52	989,97
0,84	-576,14	227,10	226,53	2,63	838,79

Note: Ecorr - corrosion potential, Jcorr - current density, eanodic - anodic slope, TC - corrosion rate and Rp is the polarisation resistance.

Table 9 shows the evolution of electrochemical parameters in relation to the controlled variation in electrolyte speed. There is a variation in corrosion potentials towards higher values, as well as an increase in current densities.

The increase in anodic gradients associated with the increase in electrolyte velocity suggests greater dissolution of metallic tones in the carbon auger sample, as a result of the aggressive and intense contact of chloride tones on the metal surface due to the dynamics of the process. The sum of these factors results in flaws in the thin,

unstable pebble formed on the metal surface. This surface has specific resistances which are modified by the dynamics of the electrolyte and which show a behaviour which is inversely proportional to the speeds investigated.
The polarisation resistance values (R_p) can be calculated using the data provided by the software (β_a β_c e i_{corr}) with the help of Stern-Geary Equation 6.8 [71]:

$$R_p = \frac{\beta_a . \beta_c}{2,303 . (\beta_a + \beta_c) . i_{corr}} \quad (6.8)$$

The corrosion rate (mm/year) can be determined by Equation 6.9 [71]:

$$TC = 0,00327 \frac{i_{corr} . EW}{\rho} \quad (6.9)$$

Where EW = equivalent weight of the auger (g), p = density of the auger (g/cm 3), icorr = corrosion current (A) and 0.00327 is the time conversion factor.
In comparison, the test carried out in the absence of flow shows a maximum Rp of 2140.03 Q, a fact related to the absence of intense diffusion at the metal-electrolyte interface. It can be seen that the speed of the electrolyte in turbulent conditions (V=0.41 $m.s^{-1}$) reduces the Rp by more than 50% compared to the system in the absence of flow. The test carried out at maximum speed (V=0.84 $m.s^{-1}$) results in maximum degradation (TC=2.63 mm/year) of the metal surface, reaching 838.79 Q of polarisation resistance. These phenomena are in line with Ferreira's explanations [52].
The explanations given by Ferreira [52] state that, in corrosion reactions, the aggressive tones must reach the metal surface by molecular diffusion as a result of a concentration gradient and the corrosion products must be displaced away from the surface. In this dynamic, the dissolved Fe+ tones move away, increasing the corrosion rate. In comparison, the thin unstable layer formed under flow conditions is thinner, more discontinuous and less resistant than the layer formed under static conditions [52] [55].

6.2.3 Correlations between electrochemical techniques

As observed through the analysis techniques applied to the samples, it is possible to make a comparison of the results obtained under the conditions studied. The aim of this section is not to exhaust the discussion or to cover all the parameters, but rather to correlate the phenomena that deal with variations that gain prominence in research involving electrochemical flow tests.
In the investigations carried out at the different speeds that guarantee turbulent flow (0.41, 0.62 and 0.84 $m.s^{-1}$ of 3% NaCl), it can be seen that in the impedance tests, the values of resistance to charge transfer reduce as the speed of the electrolyte increases. The electrolyte also modifies the structure of the electrical double layer. In the polarisation tests, a similar behaviour was observed in the same speed range.
Although it is a destructive test, compared to the impedance test, the polarisation result shows an increase in current density values, corrosion rates and a decrease in polarisation resistance values in turbulent conditions.
In general terms, the high resistance to charge transfer obtained in the impedance test implies a low corrosion rate in the polarisation tests, which in turn is inversely proportional to the polarisation resistance. This comparison between the techniques

is in line with the approaches of Domingues [86] and Ribeiro [87].

In this way, the mapping of the electrochemical system and its response to the different speeds investigated was demonstrated experimentally. It can be seen that the polarisation and impedance techniques corroborate each other, even in turbulent conditions, indicating that the speeds of 0.62 and 0.84 $m.s^{-1}$ are the most aggressive for the case studied. The high aggressiveness of these speeds can be seen in the low load transfer resistance values, high anodic inclinations and high corrosion rates (Table 8 and 9).

6.3 Part III: Tests with various oils/extracts in 3% NaCl medium under flow turbulent

At this stage, some preliminary electrochemical impedance evaluations were carried out on Babagu oil, Palm kernel oil and Mint and Limonene extracts. The aim was to assess the behaviour of these inputs in the flow condition of interest, as well as to demonstrate the progress of this research in terms of the choice of working input. The results are summarised in Figure 34.

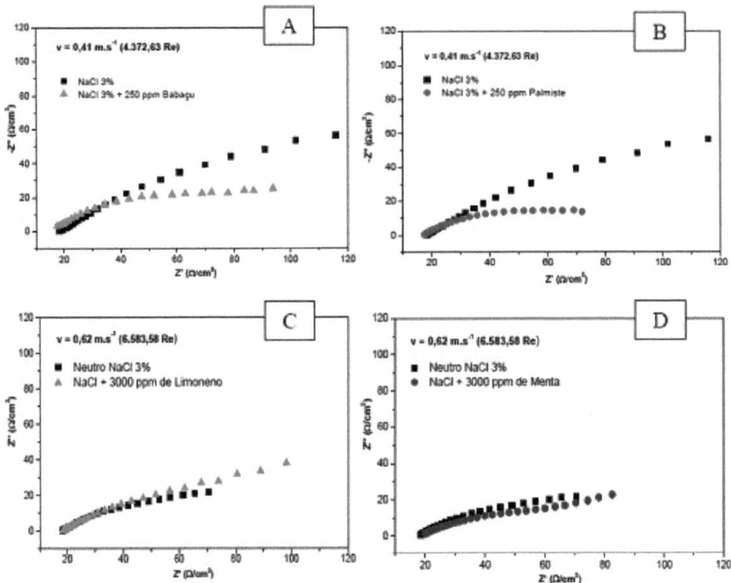

Figura 34: Nyquist spectra obtained in the presence of Babagu oil (A), Palm kernel oil (B), Limonene oil (C) and Mint oil (D) in a 3% NaCl medium under turbulent flow conditions (range 10^5 Hz to 1Hz).

Table 10 highlights the concentration used, the extraction method of the extracts/oils, the visual aspect of the solutions and some operational observations inherent to the working inputs.

Table 10: Observations on extracts/oils and working solutions in room temperature.

IFMA	Concentration	Extraction method/oil	Visual appearance of the mixture	Observation

			after stirring	
TEST STAGE I *(PALM KERNEL AND BABASSU)*	3% NaCl + 250 ppm Palmiste	Cold pressing		Presence of dispersed phases
TEST STAGE I *(PALM KERNEL AND BABASSU)*	Babassu ppm	Cold pressing		Presence of dispersed phases
TESTING STAGE II *(MINT AND LIMONENE)*	3% NaCl + 3000 ppm Mint	Steam distillation		Good interaction
TESTING STAGE II *(MINT AND LIMONENE)*	3% NaCl + 3000 ppm Limonene	Steam distillation		Good interaction

Taking into account the size of the arc in the impedance tests, the visual aspect of the working solutions and the formation of dispersed phases observed in the Babagu and Palma oils obtained by cold pressing, it was found that they did not present satisfactory results in the flow condition investigated, thus making it impossible to carry out a more comprehensive analysis of these oils as corrosion inhibitors in a system with a turbulent flow of 3% NaCl.

Since the tests carried out with the Limonene extract enabled more complete discussions, as will be seen below, and since it was clear that this extract acted as a corrosion inhibitor, it was decided to work with it in this study.

6.4 Part IV: Tests in the presence of Limonene Extract at various concentrations at 0.62 m.s-1 of 3% NaCl.

After checking the functionality of the system at various speeds, it was found that the speed of 0.62 $m.s.^{-1}$ was highly aggressive to the ago carbon sample. Based on the results of the tests with various oils/extracts, it was decided to apply the Limonene extract in various concentrations (250, 500, 1000, 3000 and 6000 ppm) in the turbulent flow regime (6583.58 Reynolds) and to evaluate the changes in obtaining the electrochemical parameters, thus analysing the efficiency of the various concentrations of this extract in this flow condition.

6.4.1 Open Circuit Potential Measurements

The aim of the potentiometry studies was to assess the ability of the extract to interact with the sample surface in turbulent flow by comparing the open circuit potential (OCP) in the absence and presence of the extract at various concentrations as a function of immersion time. Figure 35 shows the potential vs. immersion time curves in the absence and presence of the Limonene extract.

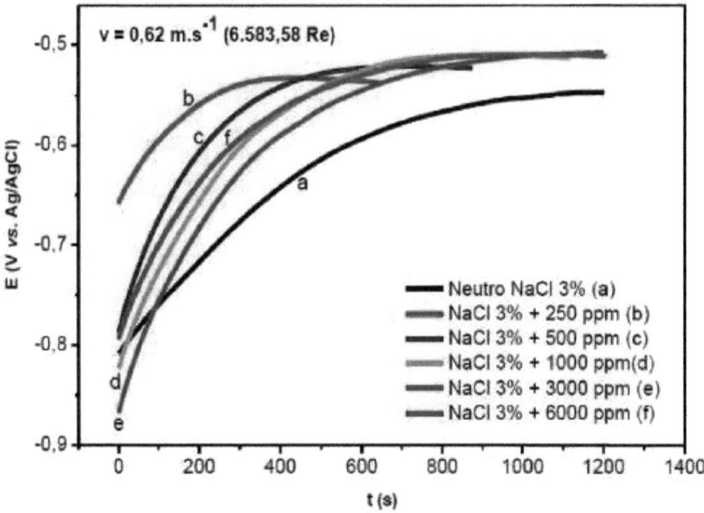

Figura 35: OCP of aug carbon in 3% NaCl medium at 0.62 m.s^{-1} with various concentrations of Limonene extract.

Table 11: Breakdown of the Open Circuit Potential curves in 3% NaCl medium at 0.62 m.s-1 with different concentrations of Limonene extract.

Concentration (ppm)	-OCP (mV)	Approximate time to start stabilisation (s)
0	546	1134
250	532	400
500	523	598
1000	512	798
3000	508	902
6000	511	915

The profile of the curves in Figure 35 shows the variations and stabilisations of the potentials under the flow conditions studied. Since this test precedes the other electrochemical tests, it is worth noting that the stabilisation achieved for each concentration ensures greater reliability of subsequent measurements.

As for the time taken for the potential to stabilise, it can be seen that the addition of the extract influences the reduction of this parameter. Specifically, the test carried out in the absence of the extract reached stabilisation in approximately 1134 s, 400 s for 250 ppm and 915 s for the maximum concentration of 6000 ppm. This behaviour suggests good interaction between the Limonene extract and the saline solution under turbulent conditions.

Evaluating the results in terms of potential variation, it can be seen that in all the concentrations analysed, the presence of the Limonene extract shifted the electrode potential towards more positive values, even in turbulent conditions. Punctually, the test carried out in the absence of the extract reached stabilisation at -546 mV and as the extract was added, the OCP took on less negative values, reaching -511 mV for the 6000 ppm concentration of Limonene extract.
According to Barros [73], Odewunmi [88] and Grosser [28], this result indicates that Limonene extract acts as an anodic inhibitor. The variations observed suggest that the anodic reactions linked to metal degradation may be being inhibited by the presence of the extract. This behaviour will be reflected in the Electrochemical Impedance and Potentiodynamic Polarisation results, as will be confirmed below.

6.4.2 Electrochemical Impedance Measurements

The Electrochemical Impedance Spectroscopy studies evaluated the effect of different concentrations of Limonene extract on obtaining electrochemical parameters in turbulent flow in 3% NaCl. Figure 36 shows the Nyquist spectra of ABNT 1020 concrete in 3% NaCl solution in the absence and presence of different concentrations of Limonene extract subjected to a flow of 0.62 $m.s^{-1}$ (6583.58 Reynolds) and Table 12 details the spectral information.

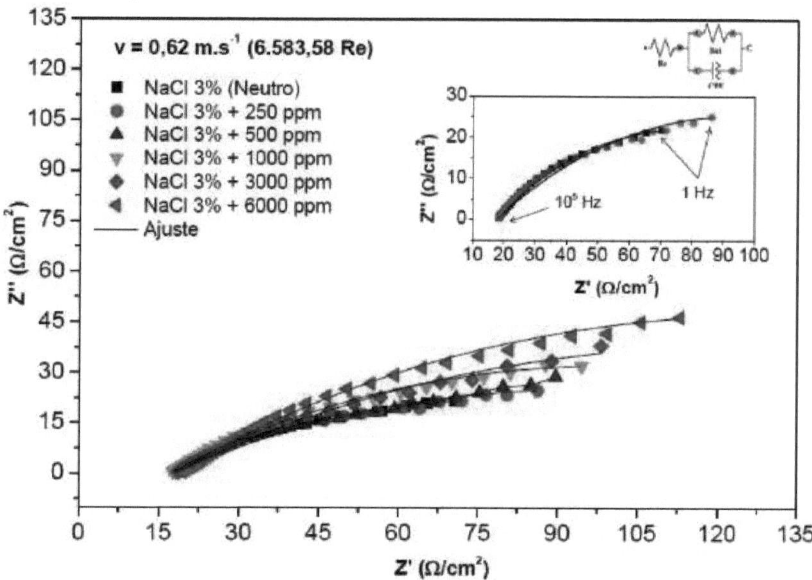

Figura 36: Open circuit potential as a function of the immersion time of carbon auger in 3% NaCl medium at 0.62 $m.s^{-1}$ with different concentrations of Limonene extract.

The spectra are presented in terms of Z' (X axis) *vs.* -Z'' (Y axis). The real component Z' represents the ohmic resistance values, while the imaginary component -Z'' represents the capacitive reactance values of the system.

Table 12: Parameters obtained from Nyquist spectra in the absence and presence of

Limonene extract at various concentrations in 3% NaCl medium at 0.62 m.s .$^{-1}$

Concentration (ppm)	Rs (Q)	Rct (Q.cm)$^{-2}$	CPE Cdl (mF.cm)$^{-2}$	n	$-X^2$	EI (%)
0	18,5	87,75	0,971	0,53	0,0104	-
250	18,6	97,6	0,490	0,52	0,0321	10
500	18,2	102	0,421	0,51	0,0496	13,97
1000	18,4	154,59	1,018	0,55	0,0094	43
3000	18,7	162,4	1,114	0,50	0,0261	45,96
6000	18,6	171,1	0,722	0,48	0,0674	48,68

Figure 36 shows the influence of the different concentrations of Limonene extract on the formation of the capacitive arcs obtained. It can be seen that the presence of Limonene extract increases the charge transfer resistance values at all the concentrations studied.

A quantitative analysis in terms of electrochemical parameters was obtained by applying the Equivalent Electrical Circuit (Rs[Rct|Q]) and is shown in Table 12. The application of this circuit to better interpret the electrochemical spectra is based on a number of investigations [42] [45] [89,90]. A

The resistance of the solution (R_s) is related to the current flow in the electrolyte. The charge transfer resistance (R_{ct}) is associated with the faradaic current density for electron transfer which, under the conditions studied, is altered by the concentration of the Limonene extract molecules at the electrode/electrolyte interface at time *t*.

The Constant Phase Element (CPE or Q) is the ideal element for analysing the data obtained, since it simulates the deviations from the ideal behaviour of a pure capacitor, taking into account the adsorption of the extract molecules on the electrode surface. This element makes it possible to analyse the capacitance of the electric double layer formed (C_{dl}), associated with the index that measures the ideality of the capacitor (*n*), which varies from 0 to 1. The inhibition efficiency was calculated using Equation 5.1 (see section 5.9).

Table 12 shows the variation in electrochemical parameters for the various concentrations investigated. It can be seen that the electrochemical measurements of solution resistance (R_s) are little influenced by the addition of the Limonene extract, since the resistance observed at the minimum concentration is the same as that found at the maximum concentration (18.6 Q).

It can be seen that increasing the concentration of the extract in the solution increases the R_{ct} values for all the concentrations studied. Figure 37 shows the influence of the different concentrations on the R_{ct} values obtained at v=0.62 m.s^{-1} of 3% NaCl (6583.58 Reynolds). The R_{ct} value in the absence of the extract is 87.75 Q/cm^2 , while the addition of 500 ppm raises this value to 102 Q/cm^2 , thus achieving an efficiency of 13.97%. The 3000 ppm concentration raises the R_{ct} value to 162.4 Q/cm^2 , whose calculated efficiency equates to 45.96%. The maximum R_{ct} value observed (171.1 Q/cm^2) occurs when the 6000 ppm concentration of the extract is added, thus providing the maximum efficiency of 48.68%.

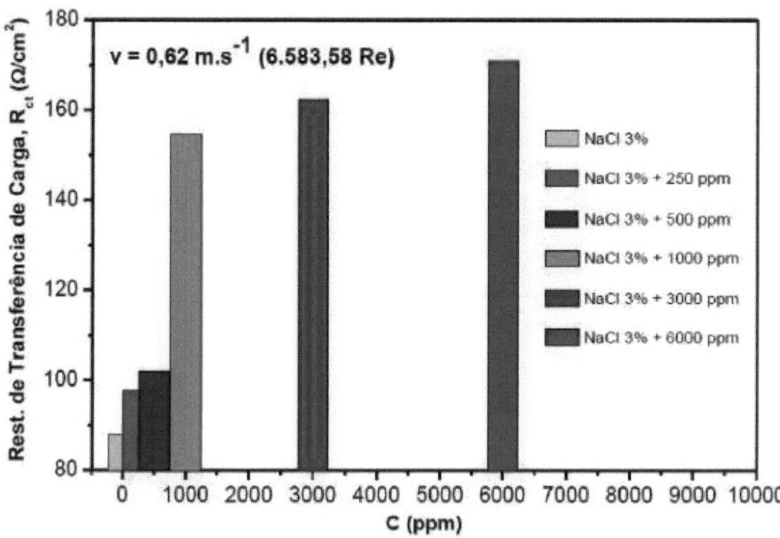

Figura 37: Effect of different concentrations of Limonene extract on obtaining Rct values in 3% NaCl medium at 0.62 m.s .$^{-1}$

Research by Jokar et al. [50], Pradityana et al. [42], Raghavendra et al. [91] and Bozorg et al. [92] associates this phenomenon with adsorption processes of the extract molecules on the metal surface. The adsorption of the molecules on the active sites of the metal surface is responsible for blocking the chloride tones that reach the interface due to the chaotic and random movements present in turbulent flow.

In this particular case, it is suggested that some Limonene molecules overcome the hydrodynamic factors and the turbulence of the process and adsorb onto specific anodic and cathodic rivers on the metal surface, thus creating a deconstructed protective monolayer at the metal/NaCl 3% interface. The efficiency values presented are directly related to the quality of the monolayer formed in this flow condition.

The capacitance of the double layer decreased as the concentration of the extract increased at 250 and 500 ppm. However, at concentrations of 1000, 3000 and 6000 ppm the capacitance values of the double layer are higher compared to the result obtained with 250 ppm of the extract. As these tests were carried out in highly aggressive flow conditions, it is possible to analyse these variations from a number of perspectives.

According to research by Pradityana et al [42], the reduction in capacitance values indicates an increase in the adsorption of inhibitor molecules on the surface of the auger at concentrations of 250 and 500 ppm. Due to the turbulent flow accompanied by higher concentrations (1000, 3000 and 6000 ppm), the increase in capacitance values may be chemically associated with the humidification of the electrolyte [93], physically related to changes in the dielectric constant of the material [42] and experimentally linked to the response mode to the frequency changes imposed in the test [93].

The heterogeneity parameter *n* decreases with increasing concentration. It is suggested that this phenomenon is associated with the adsorption of the extract and the formation of a porous layer on the metal surface [94,95].
In Table 12, x^2 is the chi-squared value showing the square of the standard deviation between the experimental points and the generated simulation. The values obtained for x^2 (0.0094 - 0.0674) indicate a good fit for the proposed circuit [96-99]. Figure 38 shows the Phase and Modulus Bode diagrams for the carbon auger obtained in the absence and presence of various concentrations of Limonene Extract in 3% NaCl solution at 0.62 m.s .$^{-1}$

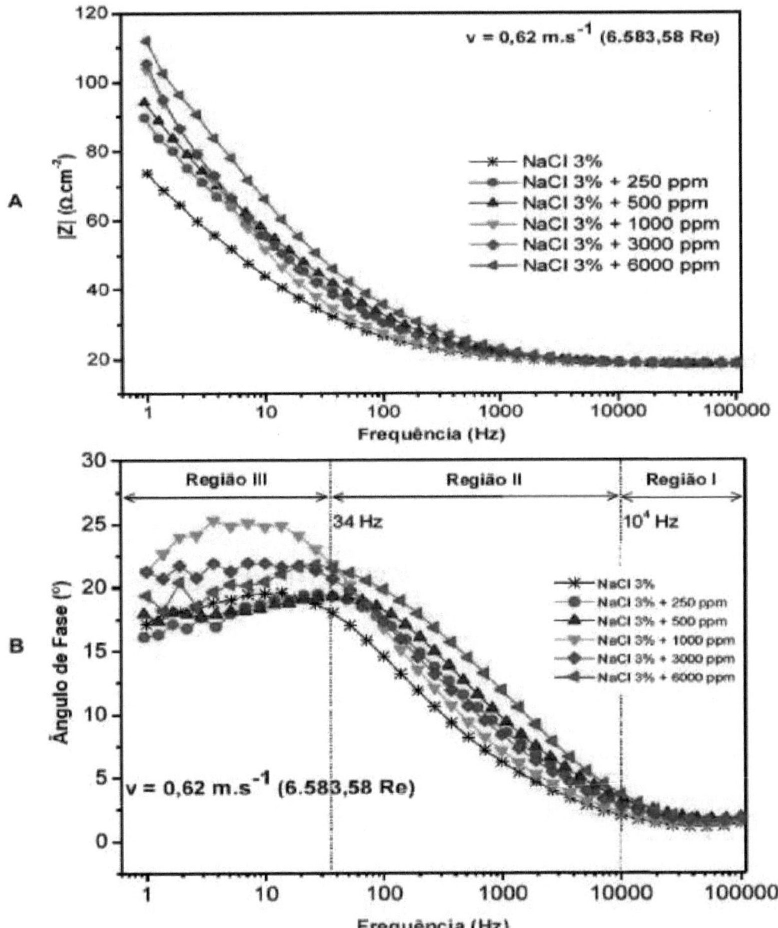

Figura 38: Bode Modulus (A) and Bode Phase (B) for the carbon auger obtained in the absence and presence of various concentrations of Limonene Extract in 3% NaCl solution at 0.62 m.s .$^{-1}$

The graph in Figure 38A shows the impedance modulus (|Z|) as a function of the frequency range (105 -1 Hz). This representation complements the Nyquist analyses, as it shows the sum of the system's capacitive and resistive contributions in the frequency range domain. The difference in the positioning of these curves corroborates the discussions based on the Nyquist spectra, confirming the increased efficiency of the Limonene extract as a green corrosion inhibitor in a system with turbulent flow.

It can be seen that between 105 and 103 Hz, the |Z| responses are approximately the same for all the concentrates studied, which indicates that the resistance of the solution persists until approximately 103 Hz. This behaviour justifies the small variation in the Rs values found (see Table 12). Graphically, the action of Limonene extract as a green corrosion inhibitor is more pronounced at frequencies below 10^3 Hz.

The graph in Figure 38B shows the variation in phase angle as a function of the experimental frequency range. In general, it can be seen that this parameter is modified, in different proportions during the test, by the addition of the Limonene extract. According to Rehim [100], the maximum phase angle for an electrochemical system adjusted to an RRQ circuit should be 90° when Rs = 0. Therefore, the low phase values (< 25°) found can be justified by the high Rs values observed in the experiments (average Rs of 18.5 Q).

The graphs show 3 regions with a specific behaviour, again indicating a different response mode to the applied frequencies. In Region I (range between 105 and 104 Hz), the phase angle is close to zero and its variation is insignificant, which may be related to the dominant effect of the solution's resistance in this frequency range.

In region II (range between 10^4 and 34 Hz), it is clear that the test carried out in the absence of the extract shows low phase angles, which may be related to an increase in the surface roughness of the electrode due to the action of aggressive tones that reach the metal surface [101]. The influence of the Limonene extract on obtaining higher phase angles in this region is noteworthy, indicating an increase in the degree of coverage and a more homogeneous distribution of the Limonene particles at the metal/electrolyte interface [101], thus contributing to an increase in inhibition efficiency.

In Region III (range between 34 and 1 Hz), there is a certain randomness in the phase angles, which suggests that the turbulence of the electrolyte plus the hydrodynamic factors compete with the electrochemical factors, thus causing this parameter to oscillate in this frequency range.

6.4.3 Potentiodynamic Polarisation Measurements

The aim of the potentiodynamic polarisation studies was to evaluate the interference of the limonene extract in obtaining the electrochemical parameters of the test. Figure 39 shows the plots of the potentiodynamic polarisation curves of the carbon auger obtained in a 3% NaCl medium at 0.62 $m.s^{-1}$ with various concentrations of Limonene extract.

The Tafel Extrapolation Method was applied to the curves to obtain the electrochemical parameters and they can be seen in Table 13. The formula for calculating the inhibitor's efficiency is based on the difference, in percentage, between the current densities in the absence and presence of the extract, following the methodology adopted in the literature [42] [69] [102], according to Equation 5.3 described above.

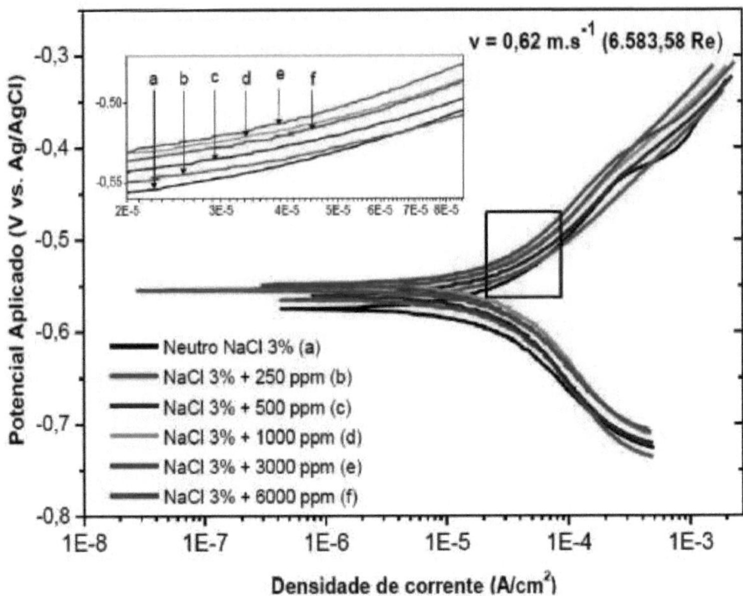

Figura 39: Potentiodynamic polarisation curves of carbon auger in 3% NaCl medium at 0.62 $m.s^{-1}$ with various concentrations of Limonene extract.

Table 13: Tafel parameters obtained from polarisation curves potentiodynamics in 3% NaCl medium at 0.62 m.s $.^{-1}$

Concentration *(ppm)*	**Ecorr.** *(mV vs. Ag/AgCl)*	**j** *corr.* *(pA/cm)²*	**Ba** (mV/dec)	**Bc** (mV/dec)	**TC** *(mm/year)*	**n%**
0	-575,28	217,69	213,34	180,01	2,52	-
250	-565,33	199,92	209,21	136,60	2,32	8,16%
500	-560,29	178,39	196,01	142,34	2,07	18%
1000	-548,32	136,9	150,41	126,26	1,58	37,32%
3000	-549,88	130,48	149,14	137,81	1,51	40,06%
6000	-554,98	126,38	133,18	136,19	1,46	42%

Table 13 presents relevant results, as it shows that the Limonene extract interferes positively in the polarisation results at all concentrations, i.e. the extract studied influenced the corrosive process, even in turbulent flow conditions. In general, the results showed that the inhibitor's efficiency increases with increasing extract concentration under the flow conditions studied (0.62 m/s). The maximum efficiency achieved in the polarisation tests was 42%, a value obtained from the addition of 6000 ppm of Limonene extract.

As for the corrosion potentials, it can be seen that the Limonene extract shifts the polarisation curves towards more anodic regions (above the curve in the absence of the extract) at all the concentrations investigated. According to Hussin [103], Wei-hua et al. [104] and Satapathy et al. [105], when the difference in the Ecorr value (ΔEcorr = Ecorr - Ecorr with inhibitor) is greater than 85 mV, the inhibitor can be classified as a

cathodic or anodic inhibitor. However, if this difference is less than 85 mV, the inhibitor can be classified as a mixed action inhibitor.
It can be analysed that, in the absence of the extract, the potential acquired is equivalent to -575.28 mV, a value which rises to a maximum potential of -548.32 mV at a concentration of 1000 ppm. At concentrations of 3000 and 6000 ppm, the potentials are reduced to -549.88 mV and -554.98 mV, respectively. Note that AEcorr does not exceed 27 mV, which indicates that the Limonene extract acts in both anodic and cathodic areas. Although in a static condition, the mixed action of Limonene extract was also observed by Hammouti et al. [69] in their research.
It can be seen that the presence of the Limonene extract reduces the current densities when compared to the test carried out in the absence of the extract. This reduction was inversely proportional to the concentrations studied. The current density in the absence of the extract is equivalent to 217.69 ^A/cm^2 and the addition of 250 ppm to the system reduces this value to 199.92 ^A/cm^2 , thus achieving the minimum efficiency found of 8.16%. The tests carried out at concentrations of 500, 1000, 3000 and 6000 ppm increased the efficiency values as described in Table 12. The maximum efficiency found was 42% in the presence of 6000 ppm of the extract, since the current density recorded was 126.38 ^A/cm .^{2}
Based on the research by Odewunmi [88], Singh et al. [106], Ji et al. [107], Boumhara et al. [108], Santos et al. [109] and Mobin et al. [110], the reduction in current density is the result of the adsorption process of the Limonene molecules on the metallic surface of the carbon auger. It is therefore suggested that some of the molecules in the extract adsorb onto certain regions of the metal surface, thus reducing current densities.
The cathodic (Pc) and anodic (Pa) inclinations and corrosion rates are also altered by the presence of the Limonene extract, as shown in Figures 40 and 41.
Figura 40: Variation of the slopes *ea* and *ec* of the polarisation curves as a function of the concentrations of Limonene extract in 3% NaCl medium.

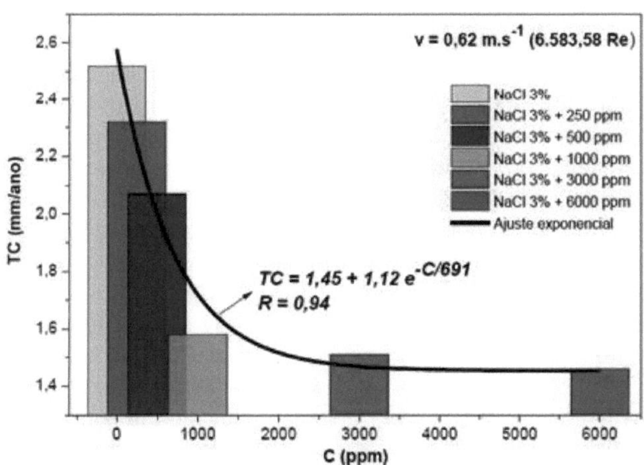

Figura 41: Effect of different concentrations of Limonene extract on the corrosion

rates of carbon auger exposed to the turbulent flow of 3% NaCl.

The differences in potentials corroborate the reduction in the Tafel slopes presented, since they prove the mixed action of the Limonene extract as a green corrosion inhibitor in the flow condition investigated.

In general, there was a drop in the slopes as the concentration of the extract increased. The anodic gradients behave significantly at concentrations of 250 to 1000 ppm, since the 1000 ppm concentration reduces the gradient by 62.93 mV/dec. The 3000 and 6000 ppm concentrations, although they do reduce the anodic gradients, show a less significant drop compared to the other concentrations.

Based on Lopez [111], this may be related to an increase in flow disturbance at these concentrations, which in turn alters the structure of the hydrodynamic boundary layer and creates a corrosion condition in which the equilibrium of the reactions is not maintained. As for the cathodic gradients, a marked reduction was observed in the presence of 250 ppm ($_{BC=136}$.60 mV/dec) when compared to the absence of the extract ($_{BC=180}$.01 mV/dec), indicating a predominant cathodic effect at this concentration.

At concentrations of 500 and 1000 ppm, there is an oscillation in this parameter which stabilises at concentrations of 3000 and 6000 ppm at values just above the cathodic inclination obtained at 1000 ppm. This phenomenon may be related to the rupture of part of the layer formed in cathodic areas due to the high surface shear stresses provided by these concentrations [111-113].

Pradityana's research [42] states that corrosion rates are affected by current densities. The test carried out in the absence of the extract showed a corrosion rate of 2.52 mm/year, which was reduced to the minimum rate found of 1.46 mm/year in the presence of 6000 ppm of the extract. In general, there was a decrease in corrosion rates at all the concentrations studied, which corroborates the efficiency values found. The experimental data was correlated using a decreasing exponential fit using Equation 6.10 with a correlation coefficient of R = 0.94.

$$TC = 1.45 + 1.12_{e}^{-c/691} \quad (6.10)$$

Where TC is the corrosion rate in mm/year, C is the concentration in ppm of the extract and the other values are constants obtained from the computer simulations. It is suggested that the exponential nature of this curve is related to the existence of a limit concentration for the action of this compound in turbulent flow conditions.

The results presented in this section corroborate the OCP and Electrochemical Impedance (Nyquist) analyses, as they demonstrate the action of the Limonene extract as a corrosion bioinhibitor. The techniques presented reveal that higher efficiencies are obtained with higher concentrations of the extract.

6.4.4 Adsorption isotherm

The action of green corrosion inhibitors is generally related to the adsorption of molecules on the metal surface. Although there is little research that discusses the particularities of this phenomenon for corrosion inhibitors exposed to turbulent conditions, the investigations of Olvera et al. [56] indicate the feasibility of this study in this flow condition, since it enables a more complete assessment of the action of the inhibitor molecules. The aim of this topic is to present a graphical analysis that provides more detailed information on the interaction between the organic molecules in the Limonene extract and the metal surface.

On protected metal surfaces, the thin protective layer formed can be expressed as the degree of surface coverage (0). This parameter is directly related to the inhibition efficiency obtained in the electrochemical impedance tests, as shown in Equation 5.2 (topic 5.9).

With this information, Figure 42 shows the plot of C/0 as a function of the C concentration in ppm. Graphically, we can see a trend line applied by means of a computational linear adjustment that correlates the concentration of the inhibitor C and the degree of coverage 0, the result of which is an almost linear correlation coefficient ($R^2 > 0.9$) of 0.9639.

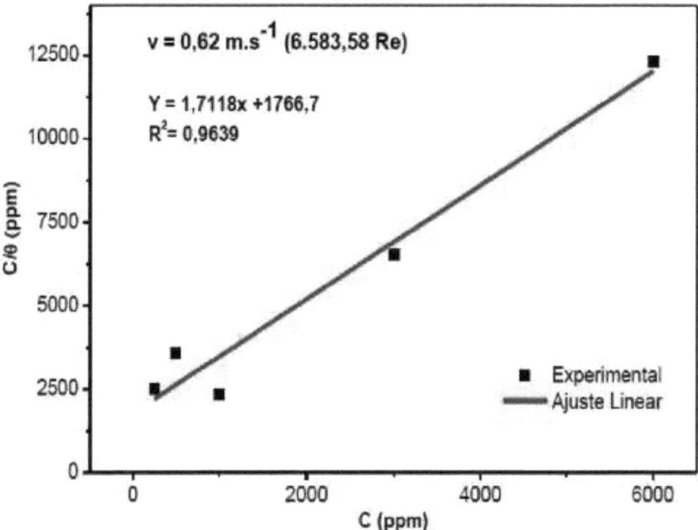

Figura 42: Langmuir adsorption isotherm for Limonene extract in 3% NaCl medium at 0.62 m/s-1.

According to some researchers [42][47,48][91,92][101], the experimental data fits the Langmuir isotherm. Based on the Langmuir model, it is proposed that some inhibitor molecules are adsorbed on the metal surface in specific, well-defined anodic and cathodic regions. All adsorption sites are energetically equivalent with perfectly flat surfaces on a microscopic scale. However, not all sites are filled by the inhibitor molecules due to the intensity of the effects provided by the turbulent flow. The adsorption of a molecule in a given stream is independent of the occupation of the neighbouring streams, i.e. it is a process without interactions between adsorbed molecules [29].

For more details, Equation 4.7 was used to calculate the adsorption constant K_{ads} (see section 4.2.1) and Equation 4.9 (see section 4.2.2) to calculate the free energy of adsorption. Based on Souza [114], with the inhibitor in extract form, the value of C_{solv} is equivalent to 1000 mg/L, *R* is the universal gas constant equivalent to 8.314 J/mol-K and T in Kelvin equivalent to 298 K. With this information and the appropriate mathematical calculations, Table 14 shows the Kads and AG°ads values:

Table 14: Kads and AGoads values for each concentration of Limonene extract in 3% NaCl medium at 0.62 m.s .$^{-1}$

Concentration (ppm)	EI (%)	Degree of Coverage (0)	W/ *e*	Kads Langmuir	- AG°ads
250	10	0,1	2477,70	0,00044	2,03
500	13,97	0,1397	3579,09	0,00032	2,82
1000	43	0,43	2325,58	0,00075	0,71
3000	45,96	0,4596	6527,41	0,00028	3,15
6000	48,68	0,4868	12325,39	0,00015	4,70

In general, when the free energy of adsorption (AG°ads) is above -20 kJ/mol, it is a physisorption process. However, if this value reaches -40 kJ/mol or less, the adsorption interaction is chemisorption [42].

Table 14 shows that adsorption occurs by physisorption, since the maximum AG°ads value obtained was -4.7 kJ/mol at a concentration of 6000 ppm. This may be associated with the effect of electrostatic interactions between the molecules, which in turn favours the formation of the thin protective layer on the metal surface. In addition, the non-polar nature of the Limonene extract can form a non-polar layer of molecules capable of resisting or blocking the polar nature of the electroKtic solution aggressive to the surface [42].

6.4.5 Mechanism of inhibition by Limonene molecules

The inhibitory effect of the Limonene extract observed on low-carbon graphite in 3% NaCl solution under the influence of turbulent flow can be explained by the process of adsorption of the extract onto the metal surface. Based on the results obtained in this study, it is possible to propose a mechanism for the adsorption of Limonene molecules under the flow conditions investigated.

Figure 43 shows a proposal for interfacial phenomena, highlighting the action of inhibitor molecules in anodic and cathodic areas of the metal surface. In this representation, 0 represents the covered surface and 1-0 the uncovered surface. In the turbulent flow condition, it is suggested that the intensity of the flow limits the inhibitory action of the Limonene extract.

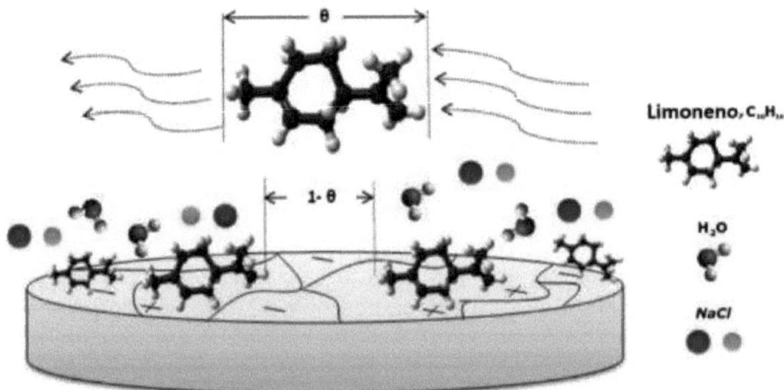

Figura 43: Proposed adsorption model of Limonene molecules subjected to turbulent flow.

It is understood that some of the molecules travelling randomly close to the metal surface resist the turbulence of the electrolyte and adsorb by means of electrostatic interactions governed by fissuring associated with the presence of conjugated double bonds or aromatic rings present in the extract, which is a characteristic shared by other Upic inhibitors [115]. According to Gerengi [116], adsorption depends on the chemical composition of the molecule, the temperature and the electrochemical potential at the metal/solution interface.

Based on Li et al [117], the H_2O molecules present in the electrolyte could also adsorb onto the metal/solution interface. However, the adsorption of organic compounds can occur by substitution between the organic compound in the aqueous phase [Org<sol>] and water molecules on the metal surface [118]. For this particular case, we suggest the reaction in Equation 6.11 presented by Gerengi [116] as fundamental for understanding the mechanism described, since it encompasses the mixed action of the Limonene extract.

$$Org_{(sol)} + xH_2O \rightarrow Org_{(ads)} + xH_2O_{(sol)} \qquad (6.11)$$

In Equation 6.11, x represents the ratio between the number of H_2O molecules replaced by the organic extract. The presence of turbulent flow makes it difficult for the inhibitor molecules to act more effectively, as this condition increases the kinetics of the reactions and the interactions that exist in the adsorption process. The increase in the inhibition efficiency of the extract indicates that the inhibitor molecules are adsorbed onto the surface of the auger with increasing concentration, thus providing an increase in the degree of coverage (0).

As such, the adsorption of molecules occurs at favoured sites, directly influenced by the distance and speed with which these molecules pass through the metal surface. It is understood that the first step in the adsorption of an organic inhibitor generally involves the replacement of one or more adsorbed H_2O molecules on the metal surface [116]. Molecules that are adsorbed in cathodic regions reduce hydrogen evolution reactions, while those that are adsorbed in anodic regions inhibit the metal dissolution process [101]. These phenomena corroborate the reduction in corrosion rates per year and the increase in efficiency found in the tests carried out.

6.5 Part V: Microscopy of the samples under the influence of flow

The aim of part V of this work is to present a visual demonstration of the effect of Limonene extract on preserving the metal surface of ABNT 1020 concrete subjected to the flow test. Figure 44 shows a photographic record comparing the metal surface exposed to the electroKtic flow of 3% NaCl at 0.62 m/s in the absence and presence of Limonene extract (6000 ppm) over a period of 6 hours.

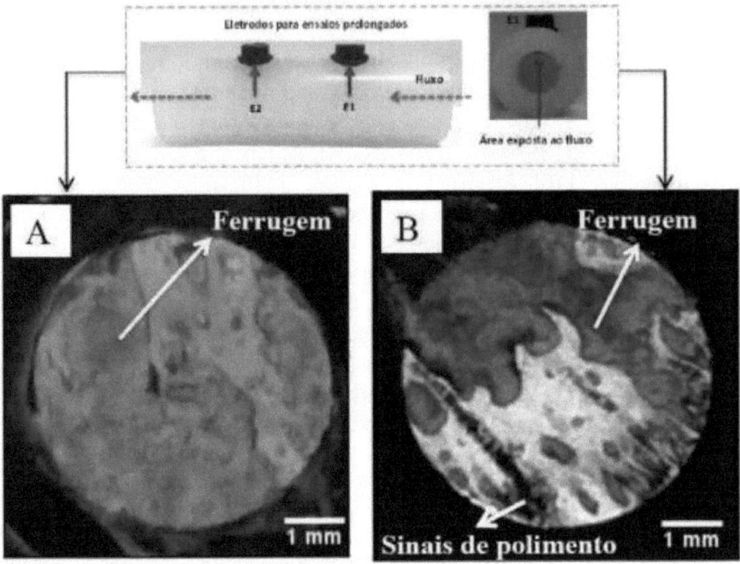

Figura 44: Photographs of the specimens exposed to the turbulent flow of 3% NaCl (0.62 m/s) for 6 hours in the absence (A) and presence of 6000 ppm Limonene (B).

We have observed in the course of this research that the action of Limonene extract as a green corrosion inhibitor is limited by the turbulent electroKtic flow in question. Thus, some aspects deserve to be highlighted in this discussion.

In general, it can be seen that in Figure 44A the surface of the test specimen in the absence of the Limonene extract has a yellowish appearance and in large proportions, representative of the corrosion products formed on this surface. In contrast, Figure 44B shows a surface less affected by this flow due to the presence of the extract, since the corrosion products observed at the end of the test are less voluminous, appear in thinner layers and are restricted to specific regions, leaving clear signs of the polishing previously carried out on the samples.

Figure 45 shows new optical microscopy records of the samples after a new battery of tests.

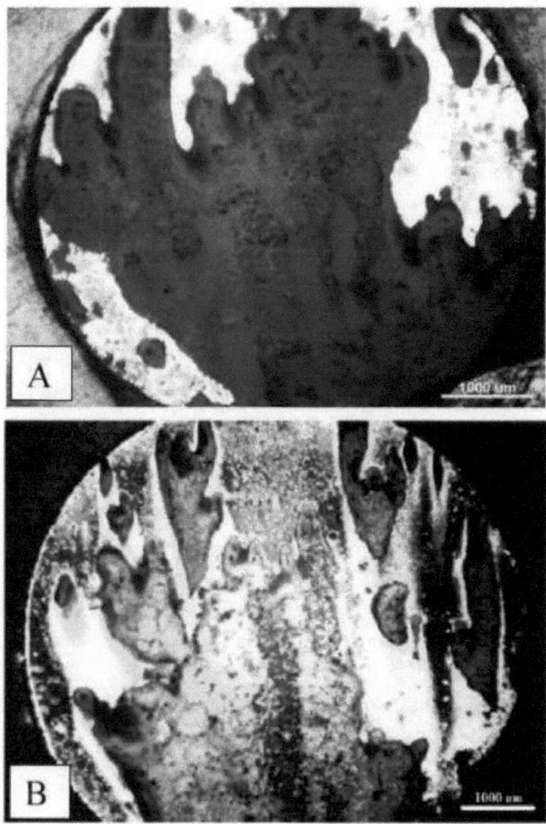

Figura 45: Microscopy of the samples after exposure for 6 hours in a turbulent flow of 3% NaCl (0.62 m/s) in the absence (A) and presence of 6000 ppm Limonene extract (B).

The morphological aspect of the sample can be seen more clearly in the presence of 6000 ppm of Limonene extract. Figure 45B clearly shows the difference between the regions most and least affected by erosive corrosion.

The differences observed from the photographic records and microscopies obtained indicate an agreement with the efficiency values found in the Electrochemical Impedance (48.68%) and Potentiodynamic Polarisation (q = 42%) tests in the presence of 6000 ppm of Limonene extract.

6.6 Part VI: Characterisation of Limonene by FTIR

Part VI of this work aims to characterise the Limonene extract in order to ascertain the bonds present and identify the possible functional groups that are promising for this extract to act as a green corrosion inhibitor.

Figure 46 shows the FTIR spectrum of the Limonene extract, showing peaks P1 to P8. The plot shows the variation in transmittance (a.u.) as a function of wave number Л (cm^{-1}). In order to better visualise the spectra, some technical resources were applied, such as: baseline alignment (BASILINE) and spectral noise smoothing (SMOOTH).

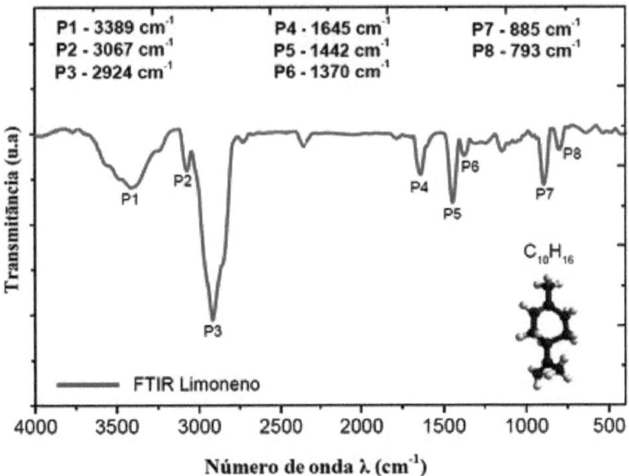

Figura 46: FTIR spectrum of Limonene extract (inhibitor).

Table 15 details the spectrum in Figure 46. There is a broad band around 3384 cm^{-1} (P1) and another band at 3067 cm^{-1} (P2). These peaks are associated with the presence of C=C-H bonds (alkene group) or =C-H bonds (alkyne group) [42]. The frequencies 2924 cm^{-1} and 793 cm^{-1} (P3 and P8) are associated with the stretching of weak C-H bonds resulting from the vibrations of alkenes present in the structure [79].

In the region near 1645 cm^{-1} (P4), the stretching of the C=C aromatic ring, at 1442 cm^{-1} (P5) the angular deformation in the C - H_2 bonds, at 1370 cm^{-1} (P6) the CH_3 group and the peak formed at frequencies near 885 cm^{-1} (P7) are associated with the R_2C = CH structure of the molecule [78] [105].

Table 15: Functional groups from the FTIR spectrum of the Limonene extract.

PICO	Wave number, Л (cm)$^{-1}$	Possible groups
P1, P2	" 3389, 3067	C=C-H; = C-H;
P3, P8	" 2924, 793	Stretching Alkenes C - H
P4	" 1645	C= C stretch
P5	" 1442	Angular deformation C - H_2
P6	" 1370	CH_3
P7	" 885	R_2C = CH

The result presented shows the presence of the bonds that identify the Limonene extract and the C=C and C-H groups observed in the P2 and P3 peaks of the spectrum stand out. According to Gerengi [116], these groups are present in the vast majority of compounds that act as green corrosion inhibitors.

7 CONCLUSIONS

The results and discussion discussed in this dissertation allow us to draw the following conclusions:

a) In relation to the construction and evaluation of the functionality of the device for electrochemical testing with flow:

- The experimental bench proved to be effective for electrochemical studies with flow control, recording different types of flow;
- The system's sensitivity to electrolyte velocity variations and the agreement between the results obtained through electrochemical techniques in turbulent flow conditions were verified.

b) Regarding potentiometric measurements (OCP):

- They showed that the Limonene extract interacted well with the 3% NaCl solution, shifting the curves towards more anodic potentials at all the concentrations investigated, thus reducing metal dissolution.

c) Regarding Electrochemical Impedance measurements:

- They showed that the resistance to charge transfer (R_{ct}) increases with the increase in the concentrations investigated;
- The 6000 ppm concentration of Limonene extract raises the R_{ct} to 171.1 Q.cm^{-2} , thus achieving an efficiency of 48.68% in the turbulent flow of 3% NaCl at 0.62 m.s ;$^{-1}$
- The inhibition power of this extract is associated with the process of adsorption of the Limonene molecule onto the metal surface.

d) Regarding Potentiodynamic Polarisation measurements:

- The corrosion rate is reduced as the concentration of Limonene extract increases;
- The efficiency obtained at a concentration of 6000 ppm was 42%. The action of this extract is linked to the existence of molecules that adsorb on the metal/electrolyte interface;
- The extract reduces current densities, raises corrosion potentials and reduces cathodic and anodic coefficients at all the concentrations investigated.

e) In relation to the action of the extract, the adsorption isotherm and the inhibition mechanism:

- Limonene extract slowed down anodic metal dissolution and cathodic hydrogen evolution reactions under the flow conditions investigated;
- The experimental points were fitted with a linear trend to the Langmuir isotherm with $R2$ = 0.9639 and the K_{ads} and $AG°_{ads}$ calculations indicate that the adsorption of the extract occurs by physisorption;
- The mechanism of inhibition at work is linked to the formation of an unconnected molecule at the metal/electrolyte interface that slows down the corrosive process.

f) With regard to the photographic records and microscopies obtained:

- They were in agreement with the efficiency values found using electrochemical techniques;

g) Regarding FTIR characterisation:

- The presence of C=C groups (close to 1645 cm^{-1}) and C-H groups (close to 2924 cm^{-1}) is promising for this extract to act as a green corrosion inhibitor.

8 SUGGESTIONS FOR FUTURE WORK

Based on the results obtained in this research, some suggestions for future work will be presented in order to complement the scope of this study:

- Further investigations were carried out at a speed of 0.84 m.$_{s-1}$ in order to check the extract's efficiency under this condition;
- Carry out new investigations with different electrolytic solutions;
- Investigate the influence of temperature on inhibition efficiency values in turbulent flow;
- Use additional modules attached to the equipment to check for possible fluctuations in pH values during the experiment;
- Perform SEM on the samples; and
- Perform computational fluid dynamics (CFD) analyses;

BIBLIOGRAPHICAL REFERENCES

[1] AL-OTAIBI, M. S. et al. **Corrosion inhibitory action of some plant extracts on the corrosion of mild steel in acidic media.** Arabian Journal of Chemistry, v. 7, n. 3, p. 340-346, 2014.

[2] OBOT, I. B.; OBI-EGBEDI, N. O.; UMOREN, S. A. **Antifungal drugs as corrosion inhibitors for aluminium in 0.1 M HCl.** Corrosion Science, v. 51, n. 8, p. 1868-1875, 2009.

[3] DARIVA, Camila G.; GALIO, Alexandre F. **Corrosion inhibitors-principles, mechanisms and applications.** Developments in corrosion protection, p. 365- 379, 2014.

[4] GARCIA, Luis Paulo; SANTOS, Aldo Ramos. **Application of corrosion inhibitors in the oil and gas industry.** 1st EPG, p. 33, 2012.

[5] NEGM, Nabel A. et al. **Gravimetric and electrochemical evaluation of environmentally friendly nonionic corrosion inhibitors for carbon steel in 1M HCl.** Corrosion Science, v. 65, p. 94-103, 2012.

[6] ABDEL-GABER, A. M. et al. **A natural extract as scale and corrosion inhibitor for steel surface in brine solution.** Desalination, v. 278, n. 1, p. 337- 342, 2011.

[7] BENSABAH, F. et al. **Chemical Composition and Inhibitory Effect of the Essential Oil from Lippia Citriodora Irrigated by Wastewater on the Alkaline Corrosion of Aluminum.** Portugaliae Electrochimica Acta, v. 32, n. 6, p. 381-393, 2014.

[8] SINGH, A. K.; MOHAPATRA, S.; PANI, B. **Corrosion inhibition effect of Aloe Vera gel: Gravimetric and electrochemical study.** Journal of Industrial and Engineering Chemistry, v. 33, p. 288-297, 2016.

[9] AFIA, L. et al. **Electrochemical Evaluation of Linseed Oil as Environmentfriendly Inhibitor for Corrosion of Steel in HCl Solution.** Portugaliae Electrochimica Acta, v. 33, n. 3, p. 137-152, 2015.

[10]LECANTE, A. et al. **Anti-corrosive properties of S. tinctoria and G. ouregou alkaloid extracts on low carbon steel.** Current Applied Physics, v. 11, n. 3, p. 714724, 2011.

[11]TRAN, Thu et al. **Investigation of the electrochemical mechanisms for acetic acid corrosion of mild steel.** Corrosion, v. 70, n. 3, p. 223-229, 2013.

[12]MAROSTICA JUNIOR, Mario Roberto; PASTORE, Glaucia Maria. **Biotransformation of limonene: a review of the main metabolic pathways.** Quimica Nova, v. 30, n. 2, p. 382-387, 2007.

[13]PORTELLA, Kleber Franke et al. **Physicochemical performance of metals and concrete structures used in electricity distribution networks: case study in Manaus region.** Quimica Nova, v. 29, n. 4, p. 724-734, 2006.

[14]CURY, A; SILVEIRA, D. Economic news. **News portal of the Globo**, margo. 2017. Available at: https://g1.globo.com/economia/noticia/pib-brasileiro-recua-36-em-2016-e-tem-pior-recessao-da-historia.ghtml. Accessed on: 31 May 2017.

[15]GENTIL, V. **Corrosion**, LTC Livro Tecnico e Cientificos SA, 6ª edition, chap. 1923. Rio de Janeiro, 2011.

[16]RAMANATHAN, Lalgudi Venkataraman. **Corrosion and its control.** 2014.
[17]CAINES, Susan; KHAN, Faisal; SHIROKOFF, John. **Analysis of pitting corrosion on steel under insulation in marine environments.** Journal of loss prevention in the process industries, v. 26, n. 6, p. 1466-1483, 2013.
[18]ROSSI, Catia Guaraciara Fernandes Teixeira. **Corrosion inhibition of AISI 1020 steel in acidic and saline media by microemulsified surfactants and nitrogenous substances.** 2007.
[19]RAJABIPOUR, Ali; MELCHERS, Robert E. **A numerical study of damage caused by combined pitting corrosion and axial stress in steel pipes.** Corrosion Science, v. 76, p. 292-301, 2013.
[20]ANJOS, Gineide Conceigao dos. **Utilisation of plant raw materials for application as corrosion inhibitors.** 2012. Master's dissertation. Federal University of Rio Grande do Norte.
[21]BREMNER, Theodore et al. **Protection of Metals in Concrete Against Corrosion**. 2001.
[22]MIGAHED, M. A. et al. **Synthesis of some novel non ionic surfactants based on tolyltriazole and evaluation their performance as corrosion inhibitors for carbon steel.** Egyptian Journal of Petroleum, v. 22, n. 1, p. 149- 160, 2013.
[23]HEGAZY, M. A. et al. **Novel cationic surfactants for corrosion inhibition of carbon steel pipelines in oil and gas wells applications.** Journal of Molecular Liquids, v. 214, p. 347-356, 2016.
[24]HEAKAL, Fakiha El-Taib; ELKHOLY, Ayman E. Gemini **Surfactants as Corrosion Inhibitors for Carbon Steel.** Journal of Molecular Liquids, 2017.
[25]OAS ENGINEERING. **Pedra do Cavalo adductor system.** Sao Paulo, [2014]. Available at:<http://www.oas.com.br/oascom/oasengenharia/realizacoes/saneamento/agua/sistemaadutor-pedra-do-cavalo/#>. Accessed on: 09 March 2017.
[26]TAWANCY, H. M.; AL-HADHRAMI, Luai M.; AL-YOUSEF, F. K. **Analysis of corroded elbow section of carbon steel piping system of an oil-gas separator vessel.** Case Studies in Engineering Failure Analysis, v. 1, n. 1, p. 6- 14, 2013.
[27]SANTOS, Erica da costa dos. **Application of waste from the malt industry as corrosion inhibitors for 1020 carbon steel in a highly corrosive environment in the oil industry** / Erica da Costa dos Santos - Rio de Janeiro: UFRJ, 2015. 107 f.; il. Dissertation (Master's) - Federal University of Rio de Janeiro, Institute of Chemistry, 2015.
[28]GROSSER, Fabiana Nogueira. **Organic corrosion inhibitors: studies with natural compounds obtained from various species of mint**. 2015.
[29]DO NASCIMENTO, Ronaldo Ferreira et al. **Adsorption: theoretical aspects and environmental applications.** Science and Technology Library, 2014.
[30]CLAUDINO, Andreia et al. **Preparation of activated carbon from peat and its use in the removal of pollutants.** 2003.
[31]ATKINS, P. W.; DE PAULA, Julio. **Physical Chemistry,** vol. 1. LTC. 9ª edition, 2012.
[32]COELHO, Gustavo Ferreira et al. **Use of adsorption techniques using agro-industrial residues in the removal of contaminants in water.** Journal of Agronomic Sciences, Umuarama, v. 3, p. 291-317.
[33]ABRACO. Brazilian Corrosion Association. **Corrosion Inhibitors.** Available

at:<http://www.abraco.org.br>. Accessed on: 27 April 2017.

[34]WILLIAM, D. CALLISTER; CALLISTER, J. R. **Materials Science and Engineering: an introduction.** Sao Paulo: LTC, 2016.

[35]DE ARAUJO NETO, Jose Amilcar Mendes. **Caatinga plant extracts as a corrosion inhibitor.** Natal-2013.

[36]YARO, Aprael S.; KHADOM, Anees A.; WAEL, Rafal K. **Apricot juice as green corrosion inhibitor of mild steel in phosphoric acid.** Alexandria Engineering Journal, v. 52, n. 1, p. 129-135, 2013.

[37]SANYAL, B. **Organic compounds as corrosion inhibitors in different environments-a review.** Progress in Organic Coatings, v. 9, n. 2, p. 165-236, 1981.

[38]GUO, Lei et al. **Theoretical insight into an empirical rule about organic corrosion inhibitors containing nitrogen, oxygen, and sulfur atoms.** Applied Surface Science, v. 406, p. 301-306, 2017.

[39]EL-HADDAD, Mahmoud N. **Chitosan as a green inhibitor for copper corrosion in acidic medium.** International journal of biological macromolecules, v. 55, p. 142-149, 2013.

[40]GAYATHRI, V. S. et al. **Green inhibitors: Anti corrosive propensity of Garcinia mangostana for Aluminum 1100.** In: Solid State Phenomena. Trans Tech Publications, 2012. p. 109-112.

[41]GUNAVATHY, N.; MURUGAVEL, S. C. **Corrosion inhibition studies of mild steel in acid medium using Musa Acuminata fruit peel extract.** Journal of Chemistry, v. 9, n. 1, p. 487-495, 2012.

[42]PRADITYANA, Atria et al. **Inhibition of Corrosion of Carbon Steel in 3.5% NaCl Solution by Myrmecodia Pendans Extract.** International Journal of Corrosion, v. 2016, 2016.

[43]VASZILCSIN, C. G. et al. **Inhibiting Corrosion Process of Carbon Steel in Sodium Chloride Aqueous Solution by Capsaicin Extract.** Series of Chemistry and Environmental Engineering, v 61 (75), 2016.

[44]YETRI, Yuli et al. **Corrosion Behavior of Environmental Friendly Inhibitor of Theobroma cacao Peels Extract for Mild Steel in NaCl 1.5 M.** Environment Asia, v. 9, n. 1, 2016.

[45]FAKIH, M.; EL-KHATEEB, A. Y.; FOUDA, A. S. **Methanolic extract of curcum plant as a green corrosion inhibitor for steel in NaCl polluted solutions.** Indian Journal of Scientific Research, v. 4, n. 2, p. 219, 2013.

[46]TOMIE, M.; V. Mieie et al. **Common sage extract as an inhibitor of steel corrosion in 3% NaCl.** International Journal of Latest Research in Science and Technology. 2014.

[47]HAIDER A.Y.; MASHHDANI, KHULOODABID SALEHAL-SAADIE et al. **Cactus as a green inhibitor for the corrosion of carbon steel in seawater.** In: Physsiiccall Chemiisstry. An Indian Journal, 2015. p. 111-120.

[48]IBRAHIM, Taleb et al. **Corrosion inhibition of mild steel by Calotropis rocera leaves extract in a CO2 saturated sodium chloride solution.** Journal of Adhesion Science and Technology, v. 30, n. 23, p. 2523-2543, 2016.

[49]AL-SULTANI Kh.F., Al-Seroury F. A., Abdulsada Sh. A. **Compare between Two Type of Eco-friendly Corrosion Inhibitors by Studying Corrosion Resistance of Carbon Steel.** Journal of Research in Chemical, v. 2, p. 79-82, 2015.

[50] JOKAR, M.; FARAHANI, T. Shahrabi; RAMEZANZADEH, B. **Electrochemical and surface characterizations of morus alba pendula leaves extract (MAPLE) as a green corrosion inhibitor for steel in 1M HCl.** Journal of the Taiwan Institute of Chemical Engineers, v. 63, p. 436-452, 2016.
[51] CHALLOUF, Hassen et al. **Origanum Majorana Extracts as Mild Steel Corrosion Green Inhibitors in Aqueous Chloride Medium.** Journal of Environmental Protection, v. 7, n. 04, p. 532, 2016.
[52] FERREIRA, Lilian Raquel Moretto. **Evaluation of flow corrosion of AISI 1020 steel in NaHCO3 solution saturated with CO2.** 2012.
[53] WHITE, Frank M. **Fluid Mechanics-6.** AMGH Editora, 2010.
[54] BRUNETTI, Franco. **Fluid mechanics.** Pearson Prentice Hall, 2007.
[55] OCHOA, Diana Maria Lopez. **Study of erosive-corrosive wear of high-nitrogen stainless steels in muddy environments.** 2007. Doctoral thesis.
[56] OLVERA-MARTINEZ, Mana Elena; MENDOZA-FLORES, Juan; GENESCA, J. **CO2 corrosion control in steel pipelines. Influence of turbulent flow on the performance of corrosion inhibitors.** Journal of Loss Prevention in the Process Industries, v. 35, p. 19-28, 2015.
[57] JING, T. I. A. N. et al. **Effect of flow velocity on corrosion behaviour of AZ91D magnesium alloy at elbow of loop system.** Transactions of Nonferrous Metals Society of China, v. 26, n. 11, p. 2857-2867, 2016.
[58] LI, Qiang; HU, Haitao; CHENG, Y. Frank. **Corrosion of pipelines in CO2-saturated oil-water emulsion flow studied by electrochemical measurements and computational fluid dynamics modelling.** Journal of Petroleum Science and Engineering, v. 147, p. 408-415, 2016
[59] LEON, LD Lopez et al. **Corrosion of Carbon Steel in a Synthetic Environment and Kerosene on a System with Continuous Flow.** Int. J. Electrochem. Sci, v. 10, p. 7582-7595, 2015.
[60] BURDOCK, G. A.; **Fenaroli's Handbook of Flavour Ingredients,** 3rd ed., CRC: Boca Raton, 1995.
[61] BARROZO, Ana Carolina Faria; DOS SANTOS, Daniele Bueno. **Extraction and Purification of Limonene Part I.**
[62] SCIENCEMA WIKI. **Molecular structure of Limonene.** Available at: http://sciencemadness.wikia.com/wiki/File:Limonene-3D-balls.png. Accessed April 2018.
[63] CORAZZA, M.; RODRIGUES, D. G.; NOZAKI, J. **Preparation and characterisation of orange wine.** Chem. Nova, v. 24, n. 4, p. 449-452, 2001.
[64] NONINO, Eliseu A. **Where is the citrus industry going? Perfumer & flavorist,** v. 22, n. 2, p. 53-58, 1997.
[65] BERGER, R. G.; Krings, U.; Zorn, H. **In Food Flavour Technology;** Taylor A. J., ed.; 2002, ch. 3. 7. available at: http://www.abecitrus.com.br, accessed in July 2018.
[66] GELB, Michael H. et al. **The inhibition of protein prenyltransferases by oxygenated metabolites of limonene and perillyl alcohol.** Cancer letters, v. 91, n. 2, p. 169-175, 1995.
[67] KAJI, Itaru et al. **Inhibition by D-limonene of experimental hepatocarcinogenesis in Sprague-Dawley rats does not involve p21^{ras} plasma membrane association.** International journal of cancer, v. 93, n. 3, p. 441-444,

2001.
[68]SOUZA, M. C. et al. **Evaluation of anti-inflammatory activity of essential oils from two Asteraceae species.** Die Pharmazie-An International Journal of Pharmaceutical Sciences, v. 58, n. 8, p. 582-586, 2003.
[69]HAMMOUTI, B. **Limonene as green inhibitor for steel corrosion in hydrochloric acid solutions.** Acta physico-chimica sinica, v. 25, n. 7, p. 12541258, 2009.
[70]MANSSOURI, M. et al. **Adsorption proprieties and inhibition of mild steel corrosion in HCl solution by the essential oil from fruit of *Moroccan Ammodaucus leucotrichus*.** J. Mater. Environ. Sci, v. 6, n. 3, p. 631-646, 2015.
[71]WOLYNEC, S. **Electrochemical Techniques in Corrosion.** Sao Paulo: Edusp, 2013.
[72]GONQALVES, Joana Carla Soares; BODE, Klaus. **Environmental education.** Oficina de Textos, 2015.
[73]BARROS, Iuri Bezerra de et al. **The inhibitory action of *Bauhinia purpurea* extracts on the corrosion of carbon steel in sulfuric acid medium.** Materials Research, n. AHEAD, p. 0-0, 2016.
[74]ROJAS, Y. A. H. **Electrochemical study and characterisation of CO2 corrosion product layers in API5L X80 and API 5CT P110 steel**. 2014. Doctoral thesis. PUC-Rio.
[75]GOMES, Edelize Angelica. **Comparative analysis of the corrosion resistance of carbon steel in field tests and electrochemical tests.** 2013.
[76]BOTT, Ivani de S. **Evaluation of the corrosion behaviour of API 5LX70 and API 5LX80 steel in CO2-containing media using surface analysis and electrochemical techniques.** 2006. Doctoral thesis. PUC-Rio.
[77]STUART, Barbara. **Infrared spectroscopy.** John Wiley & Sons, Inc., 2005.
[78]MARTINS, Fernando Jose da Costa Dias. **Controlled release systems for essential oils: evaluation by infrared spectroscopy**. 2014.
[79]SILVERSTEIN, Robert M. et al. **Spectrometric identification of organic compounds.** 6ª edigao, Rio de Janeiro: LTC - Livros Tecnicos e Cientificos Editora S.A., 2000.
[80]HAACK, Micheli de Souza. **Analysis of materials by infrared spectroscopy within the quality management system according to ABNT NBR ISO/IEC 17025**. 2010.
[81]POORQASEMI, E. et al. **"Investigating accuracy of the Tafel extrapolation method in HCl solutions"**, Corrosion Science, v. 51, n. 5, p. 1043-1054, 2009.
[82]STANDARD, A. S. T. M. **G59-97: Standard Test Method for Conducting Potentiodynamic Polarisation Resistance Measurements**. West Conshohocken, PA: ASTM International, 2009.
[83]BARROW, Bruce et al. **American National Standard for Metric Practice (Institute of Electrical and Electronics Engineers**, Inc., New York, and ASTM International, West Conshohocken, PA, 2011). IEEE/ASTM SI-10.
[84]ASTM, G. 184-06 **"Standard Practice for Evaluating and Qualifying Oilfield and Refinery Corrosion Inhibitors Using Rotating Cage".** Current edition approved on Jan15, 2006.
[85]ASTM E1252: **"Standard Practice for General Techniques for Obtaining**

Infrared Spectra for Qualitative Analysis". United States, 2007.

[86] DOMINGUES, Hellen Cristiane Nunes. **"Study of the electrochemical behaviour of carbon auger in sodium bicarbonate and carbon dioxide gas using the electrochemical noise technique",** 2011. Master's thesis.

[87] RIBEIRO, Daniel Veras; SOUZA, Carlos Alberto Caldas; ABRANTES, JCC. **Use of Electrochemical Impedance Spectroscopy (EIE) to monitor corrosion in reinforced concrete.** IBRACON Journal of Structures and Materials , v. 529-546, 2015.

[88] ODEWUNMI, N. A.; UMOREN, S. A.; GASEM, Z. M. **Watermelon waste products as green corrosion inhibitors for mild steel in HCl solution.** Journal of Environmental Chemical Engineering, v. 3, n. 1, p. 286- 296, 2015.

[89] CHRAIBI, M. et al. **Antibacterial activity and corrosion inhibition of mild steel in 1.0 M hydrochloric acid solution by M. piperita and M. pulegium essential oils.** Journal of Materials and Environmental Sciences, v. 8, n. 3, p. 972981, 2017.

[90] SHAHIDI ZANDI, Mehdi; GOLESTANI, Ghasem; GHOLAMHOSSEINZADEH, Mohammad Reza. **Mentha spicata L. extract as a green corrosion inhibitor for carbon steel in HCl solution.** Physical Chemistry Research, v. 5, n. 2, p. 293-307, 2017.

[91] RAGHAVENDRA, N.; BHAT, J. Ishwara. **Natural products for material protection: an interesting and efficacious anticorrosive property of dry arecanut seed extract at electrode (aluminium)-electrolyte (hydrochloric acid) interface.** Journal of Bio-and Tribo-Corrosion, v. 2, n. 4, p. 21, 2016.

[92] BOZORG, Mansoor et al. ***Myrtus communis*** **as green inhibitor of copper corrosion in sulfuric acid.** Industrial & Engineering Chemistry Research, v. 53, n. 11, p. 4295-4303, 2014.

[93] KURZWEIL, P.; FISCHLE, H.-J. **A new monitoring method for electrochemical aggregates by impedance spectroscopy.** Journal of Power Sources, v. 127, n. 1-2, p. 331-340, 2004.

[94] POPOVA, A.; CHRISTOV, M. **Evaluation of impedance measurements on mild steel corrosion in acid media in the presence of heterocyclic compounds.** Corrosion Science, v. 48, n. 10, p. 3208-3221, 2006.

[95] GERENGI, Husnu et al. **Evaluation of corrosion inhibition of brass-118 in artificial seawater by benzotriazole using Dynamic EIS.** Corrosion Science, v. 51, n. 11, p. 2573-2579, 2009.

[96] OLASUNKANMI, Lukman O. et al. **Some quinoxalin-6-yl derivatives as corrosion inhibitors for mild steel in hydrochloric acid: experimental and theoretical studies.** The Journal of Physical Chemistry C, v. 119, n. 28, p. 1600416019, 2015.

[97] SALEHIFAR, Nahideh et al. **Electrochemical study of supercapacitor performance of polypyrrole ternary nanocomposite electrode by fast Fourier transform continuous cyclic voltammetry.** RSC Advances, v. 5, n. 116, p. 9613096137, 2015.

[98] ROY, Pialee et al. **Adsorption behaviour of gluten hydrolysate on mild steel in 1 M HCl and its role as a green corrosion inhibitor.** RSC Advances, v. 5, n. 75, p. 61170-61178, 2015.

[99]ALKHATHLAN, H. Z. et al. **Anticorrosive assay-guided isolation of active phytoconstituents from *Anthemis pseudocotula* extracts and a detailed study of their effects on the corrosion of mild steel in acidic media.** RSC Advances, v. 5, n. 67, p. 54283-54292, 2015.
[100] REHIM, Sayed S. Abdel et al. **On the corrosion inhibition of low carbon steel in concentrated sulphuric acid solutions. Part I: Chemical and electrochemical (AC and DC) studies.** Corrosion Science, v. 50, n. 8, p. 22582271, 2008.
[101] KRISHNEGOWDA, Pavithra M. et al. **Acalypha torta leaf extract as green corrosion inhibitor for mild steel in hydrochloric acid solution.** Industrial & Engineering Chemistry Research, v. 52, n. 2, p. 722-728, 2013.
[102] SHIVAKUMAR, S. S.; MOHANA, K. N. **Studies on the inhibitive performance of Cinnamomum zeylanicum extracts on the corrosion of mild steel in hydrochloric acid and sulphuric acid media.** Journal of Materials and Environmental Science, v. 4, n. 3, Cop, p. 448-459, 2013.
[103] M. H. HUSSIN, M. Hazwan; KASSIM, M. Jain. **The corrosion inhibition and adsorption behaviour of Uncaria gambir extract on mild steel in 1 M HCl.** Materials Chemistry and Physics, v. 125, n. 3, p. 461-468, 2011.
[104] LI, Wei-hua et al. **Some new triazole derivatives as inhibitors for mild steel corrosion in acidic medium.** Journal of Applied Electrochemistry, v. 38, n. 3, p. 289-295, 2008.
[105] SATAPATHY, A. K. et al. **Corrosion inhibition by *Justicia gendarussa* plant extract in hydrochloric acid solution.** Corrosion science, v. 51, n. 12, p. 2848-2856, 2009.
[106] SINGH, M. Ramananda; GUPTA, Prachi; GUPTA, Kaushal. **The litchi (*Litchi Chinensis*) peels extract as a potential green inhibitor in prevention of corrosion of mild steel in 0.5 M H2SO4 solution.** Arabian Journal of Chemistry, 2015.
[107] JI, Gopal et al. ***Musa paradisica* peel extract as green corrosion inhibitor for mild steel in HCl solution.** Corrosion Science, v. 90, p. 107-117, 2015.
[108] BOUMHARA, K. et al. **Artemisia Mesatlantica essential oil as green inhibitor for carbon steel corrosion in 1 M HCl solution: Electrochemical and XPS investigations.** Journal of Industrial and Engineering Chemistry, v. 29, p. 146155, 2015.
[109] SANTOS, Jorge Edson Pinheiro dos et al. **Study of the Corrosion Inhibition of Carbon Steel in acidic chloride by Mangiferin.** Materia (Rio de Janeiro), v. 21, n. 4, p. 1045-1053, 2016.
[110] MOBIN, Mohammad; RIZVI, Marziya. **Polysaccharide from Plantago as a green corrosion inhibitor for carbon steel in 1M HCl solution.** Carbohydrate Polymers, v. 160, p. 172-183, 2017.
[111] LOPEZ, Diana; FALLEIROS, Neusa Alonso; TSCHIPTSCHIN, Andre Paulo. **Effect of electrolyte flow on the corrosion resistance of AISI 304L austenitic stainless steel.** Technology in Metallurgy, Materials and Mining, v. 5, n. 4, p. 198-203, 2013.
[112] KURNIAWAN, Fredy; MADURANI, Kartika A. **Electrochemical and optical microscopy study of red pepper seed oil corrosion inhibition by self-**

assembled monolayers (SAM) on 304 SS. Progress in Organic Coatings, v. 88, p. 256-262, 2015.

[113] QURAISHI, M. A. et al. **Green approach to corrosion inhibition of mild steel in hydrochloric acid and sulphuric acid solutions by the extract of *Murraya koenigii leaves.*** Materials chemistry and Physics, v. 122, n. 1, p. 114122, 2010.

[114] DE SOUZA, Fernando Silvio; SPINELLI, Almir. **Caffeic acid as a green corrosion inhibitor for carbon auger.** Corrosion Science, v. 51, n. 3, pag. 642649, 2009.

[115] SINGH, Ambrish; SINGH, V. K.; QURAISHI, M. A. **Aqueous extract of Kalmegh (*Andrographis paniculata*) leaves as green inhibitor for mild steel in hydrochloric acid solution.** International Journal of Corrosion, v. 2010, 2010.

[116] GERENGI, Husnu; SAHIN, Halil Ibrahim. ***Schinopsis lorentzii* extract as a green corrosion inhibitor for low carbon steel in 1 M HCl solution.** Industrial & Engineering Chemistry Research, v. 51, n. 2, p. 780-787, 2011.

[117] LI, Xianghong et al. **Inhibition effect of 6-benzylaminopurine on the corrosion of cold rolled steel in H2SO4 solution.** Corrosion Science, v. 51, n. 3, p. 620-634, 2009.

[118] §AHIN, M.; BILGIC, S.; YILMAZ, H. **The inhibition effects of some cyclic nitrogen compounds on the corrosion of the steel in NaCl mediums.** Applied Surface Science, v. 195, n. 1-4, p. 1-7, 2002.

ANNEX

Certificate of Analysis for Orange D' Limonene extract issued by the Bauru Distillery.

CERTIFICADO DE ANÁLISE

PRODUTO: D' LIMONENO DE LARANJA

DATA FABRICAÇÃO: 11/07/2017 DATA VALIDADE: 11/07/2019

NOME CIENTÍFICO: *Limonene*

LOTE Nº: DBISU-DLMNN2133081

CAS: 5989-27-5

COR:	INCOLOR/TRANSPARENTE
ODOR	CARACTERÍSTICO DE LARANJA
ASPECTO	LIQUIDO

	ESPECIFICAÇÃO	RESULTADO
TEOR DE ALDEIDO (%)	0,37 - 1,50	0,80
DENSIDADE	0,838 - 0,843	0,840
INDICE DE REFRAÇÃO	1,470 - 1,475	1,4710

ESTOCAGEM: MANTER O RECIPIENTE CHEIO E BEM LACRADO, LIVRE DE LUZ E DA PRESENÇA DE AR, EM LOCAL FRESCO E VENTILADO.

EMBALAGENS DISPONIVEIS: BOMBONA PLASTICA DE 05 LT / 20 LT / 27 LT / 45LT.

SHELF LIFE: SE MANTIDO EM BOAS CONDIÇÕES DE ARMAZENAGEM E MANUSEIO 24 MESES APÓS A FABRICAÇÃO.

Emitido eletronicamente, dispensa assinatura.

Printed by Books on Demand GmbH, Norderstedt / Germany